Practical
Beekeeping
in New Zealand

Fifth edition

Practical
Beekeeping
in New Zealand

Fifth edition

Andrew Matheson and Murray Reid

EXISLE
PUBLISHING

This edition published 2018
Previous editions published 1984, 1993, 1997, 2011

Exisle Publishing Ltd
226 High Street, Dunedin 9016, New Zealand
PO Box 864, Chatswood, NSW 2057, Australia
www.exislepublishing.com

A CiP record for this book is available from the
National Library of New Zealand.

ISBN 978-1-77559-362-1

Text design and production by Janine Brougham
Cover design by Nick Turzynski, redinc
Printed in China

6 8 10 9 7 5

This book uses paper sourced under ISO 14001 guidelines from
well-managed forests and other controlled sources.

CONTENTS

Acknowledgements   6

Preface   7

1   Is beekeeping for me?   8

2   Getting started with bees   12

3   Hive equipment   27

4   Honey bee biology   45

5   How to handle bees   65

6   Nectar and pollen sources   72

7   Colony management   83

8   Feeding bees   98

9   Swarms and swarm prevention   108

10   Dividing and uniting colonies   115

11   Moving hives   120

12   Queen bees   125

13   Parasites, diseases and pests   144

14   Harvesting honey   186

15   Honey composition and properties   193

16   Extracting and processing honey   201

17   Comb honey   211

18   Beeswax   217

19   Pollination   226

20   Laws affecting beekeeping   232

21   Beekeeping in New Zealand   243

Further sources of information   253

Glossary   261

Handy tips   273

Index   280

Acknowledgements

Two long-time colleagues critically reviewed the manuscript at the draft stage — Cliff Van Eaton scrutinised the entire text, and Mark Goodwin commented on the parasites, diseases and pests chapter. We are grateful for their contributions, though of course any errors or incorrect emphases in the final publication remain ours.

We are grateful to the following individuals and organisations for their generosity in allowing the use of illustrations in this book:
- Peter Bray, Airborne Honey: Figs 6.1, 13.35, 15.2, 15.3, 17.6.
- Trevor Bryant: Figs 4.2, 14.6.
- Collection of the late Vince Cook: Fig. 2.9.
- Crown Copyright, courtesy of the Food and Environment Research Agency National Bee Unit: Figs 4.6, 4.7, 13.4, 13.17, 13.28–13.30.
- Barry Donovan: Fig. 4.18.
- Jane Lorimer, Hillcrest Apiaries: Fig. 12.11.
- Jody Mitchell, Kaimai Range Honey: Figs 1.2, 2.4, 2.5, 4.11, 4.12, 11.2, 18.3.
- Mark Goodwin, New Zealand Institute of Plant and Food Research: Figs 2.8, 3.10, 3.16, 4.5, 4.8, 6.2, 12.8, 13.1, 13.3, 13.5, 13.6, 13.8–13.14, 13.16, 13.18–13.22, 13.24–27, 20.2.
- Frank and Mary-Ann Lindsay, Lindsays' Apiaries: Fig. 3.14.
- Ministry of Agriculture and Forestry: Figs 20.4, 20.5, 20.6.
- Operationbee.org: Fig. 4.16.
- Pdphoto.org: Figs 4.13, 4.14, 19.1.
- Collection of the late Herman Van Puffelen: Figs 4.3, 4.17.
- John and Helen Wright, South Auckland Apiaries: Fig. 17.1.

Preface

Nearly three decades ago *Practical Beekeeping in New Zealand* was written to fill the need for a comprehensive and up-to-date handbook on New Zealand apiculture. It enjoyed wide success, not only in New Zealand but to a surprising degree with beekeepers in other temperate climates who were looking for a practical manual.

Over the years the title was revised twice to take account of changes in scientific knowledge and beekeeping practice. We then collaborated to produce a fourth edition that was completely revised and reordered, especially to take account of the dramatic and irrevocable changes brought about by the presence in New Zealand of the parasitic mite varroa. That edition is now updated to full colour.

Our aim in writing this book has been to give practical guidance on hive management, and to offer an insight into New Zealand beekeeping practices. For those with no beekeeping experience, it will explain what is involved in becoming a beekeeper and describe the industry in New Zealand. If you already keep bees, the book will deal with all the steps involved in managing colonies throughout the year and how to handle hive products. For overseas readers, it will provide an insight into New Zealand apiculture.

Any book that gives a recipe for beekeeping is doomed to failure. A honey bee colony is a complex and dynamic organisation, and is subject to the normal variations of natural systems rather than to a calendar or rule book. It is essential to understand bees before you can keep them, and all through the book we have sought to provide the reasons for the techniques under discussion.

Practical Beekeeping in New Zealand has been written from our decades of experience as beekeeping tutors and apicultural consultants, and is a result of discussions with many people who wanted to find out what they should be doing with their bees, or how they could get involved with beekeeping. The book has benefited from the multitude of questions about the 'whys' of beekeeping, as well as the 'hows'.

While writing the latest edition of the book we received valuable assistance from colleagues who reviewed the manuscript, and generous provision of illustrations, which we have acknowledged separately. The book has also benefited from the ideas, suggestions and questions of colleagues in New Zealand and other countries, with whom we've had the privilege of working during our beekeeping careers.

Andrew Matheson
Murray Reid

1 Is beekeeping for me?

Beekeeping can be an immensely rewarding hobby, a source of guilt if you cannot give colonies the necessary attention at the right time, an expensive pastime, or a useful secondary source of income. Before getting into beekeeping you should think about what you want from it, and what you are prepared to put in.

People start keeping bees for many different reasons. Some simply enjoy honey and want to produce enough for their own needs. For many, a general interest in the honey bee or ecology prompts them to acquire hives. Others have economic reasons, and are attracted by the ideas of 'free' honey or pollination, or a profitable sideline business.

BEFORE YOU START

Here are a few things you should think about before deciding whether to become involved in beekeeping.

Allergic reactions

It is impossible to keep bees without being stung. Even if you always wear a complete set of protective clothing, you will get stung from time to time. Being stung is always painful, and localised swelling and itching is common. Most people do become accustomed to frequent stings, and eventually experience only minor swelling and itching.

A few people, though, don't adjust in this way, and their reaction to stings may become increasingly severe — involving swelling, itchiness and a rash away from the sting site, or even difficulty in breathing. If you have these allergic reactions, or have never been stung before, consult a doctor before deciding to take up beekeeping.

You also need to consider family members. Bees can remain in vehicles or come into buildings and sting the unwary, and venom is brought home on clothing and gloves.

Diseases and chemicals

Now that the parasitic mite varroa is well established in New Zealand, you have to actively keep bees and not merely have them. Honey bees cannot survive for long in the presence of varroa without human intervention, and this usually means applying some form of mite-control chemical several times each year. Untreated honey bee colonies will die.

Heavy work

Beekeeping is heavy work and requires good physical fitness. Boxes of honey may weigh up to 40 kg when full (the same as a bag of cement). When lifting them you will be wearing cumbersome protective clothing, often lifting the boxes from the ground and using your fingertips in relatively small handholds. The heaviest lifting is also done at the hottest time of year. However, using three-quarter depth boxes reduces the weight a beekeeper must lift.

Shifting hives can be even more difficult and is generally a two-person operation, as a two-storey, full-depth hive can weigh 70 kg or more, depending on how much honey it contains.

Timing

You must visit your hives regularly if they are to stay healthy and productive, and if you are to remain confident and satisfied with your beekeeping. The biggest workload is in spring, and it is also in spring that fine weather seldom seems to coincide with weekends. Certain critical beekeeping tasks must be carried out when colony conditions dictate, not at the beekeeper's convenience.

Eyesight

Good eyesight is needed for finding queens, looking for eggs, and diagnosing brood diseases. If you need glasses to see things close up or in fine detail, be sure to wear them when working with hives.

Economics

Beekeeping equipment, sugar for feed and chemicals for varroa control are expensive, and keeping just a few hives can be a costly hobby. But one or two dozen hives should pay for themselves, and with anything over about 25 hives beekeeping starts to become more than a hobby.

Now that varroa is established in New Zealand, full-time beekeepers find they can't run as many hives per labour unit as they once did. Fortunately an increase in honey prices, especially manuka, and in pollination fees, has compensated somewhat for the extra costs and time involved in managing varroa. Now in the age of varroa commercial beekeepers generally run 400–500 hives per labour unit on average, whereas before varroa it was 600–800 or even more.

Red tape

There are some restrictions on keeping bees and selling bee products, as well as some annual compliance costs, but generally speaking beekeeping is hassle-free provided you are not creating a nuisance to others.

The main things you need to know are that:
- you will require the permission of the landowner before siting your hives
- the places where bees are kept (apiaries) need to be registered with the management agency for the American foulbrood pest management strategy
- there is an annual registration fee of $20 per beekeeper and up to $15 per apiary
- you must control American foulbrood disease and report on this
- some cities have restrictions on keeping bees and may charge a fee for registering your apiary, and some may require you to get the written permission of your neighbours to keep bees
- there are controls on processing bee products if you want to sell them
- there are new controls to prevent toxic honey from being produced and offered for sale for human consumption.

These and other regulations are covered in Chapter 20.

HOW TO GET STARTED

The best way to test your liking for bee-keeping is to gain practical experience before you get any hives of your own — either with an individual beekeeper or by joining a local beekeeping club. To start full of enthusiasm and then give up through loss of interest can be expensive. If you are not prepared to look after hives properly, don't get any. Varroa will kill any colonies that are not managed on a regular basis, and neglected hives are a nuisance to the public and a potential source of bee diseases.

Fig. 1.1 Learn from another beekeeper before you start beekeeping.

WHEN TO START

If you want an early return through a honey crop, spring is the best time to start keeping bees. In most areas of New Zealand the most suitable months are September to November, when early nectar and pollen sources are in flower and colonies have time to build up in strength before the summer honey flow. This is, though, a period when colonies need careful management.

It is easier from a management point of view to begin beekeeping during the summer, though unless you start with full-strength hives you may not produce any surplus honey until the following season. Beginning beekeeping in autumn means that careful management will be needed and no surplus honey will be produced until the following season. Colonies acquired in winter are difficult to check for brood diseases.

IF YOU DO GET INVOLVED IN BEEKEEPING

Once you become a beekeeper you will be joining a community of enthusiasts, and you will be part of one of New Zealand's important primary industries. Honey, beeswax, pollen, bees and minor bee products are exported to different parts of the world. Within New Zealand, honey is enjoyed as a valuable food in its own right, as well as being a useful substitute for imported

Fig. 1.2 While we don't recommend inspecting bees without wearing a veil, this does show bees can be very docile.

sugar and sugar-based products. Some honeys like manuka have gained a reputation for their antibiotic properties, while others have anti-oxidant and anti-inflammatory characteristics. Honey is now widely used in the food manufacturing and cosmetic industries.

You will also be part of beekeeping's most valuable service, which is to pollinate New Zealand's horticultural and agricultural crops and pasture legumes, a function worth billions of dollars annually.

Fig. 1.3 A well-tended apiary in a home garden.

2 Getting started with bees

Once you've decided to begin beekeeping, you will first have to kit yourself out with protective clothing. To begin with you might borrow some gear, but once you decide to get hives of your own you will need to buy equipment and make some important decisions. This initial phase of beekeeping can be expensive, but it's worth getting the right equipment as good-quality gear will last well if looked after.

PROTECTIVE EQUIPMENT

If you want to feel confident when handling bees it is important that you wear the right protective clothing. There is no shame attached to being a 'well-dressed beekeeper' if it will make you a better beekeeper. Some of the gear can be dispensed with as you gain experience.

Protective equipment can be purchased from bee equipment stockists or made at home. It must be adequate for the job though, as poorly made clothing will not help you to gain confidence. The minimum protective gear for a beekeeper is: smoker, veil, overalls, boots and — to begin with at least — gloves.

Smoker

This is probably the most important piece of beekeeper's protective equipment, and is used to calm bees so they are less likely to sting. How to use a smoker is described in Chapter 5.

Smokers come in various shapes and sizes, mostly with an 80 mm or 100 mm diameter smoke chamber. Preferences vary, but smokers with a 100 mm diameter chamber are more common and are probably easier to keep alight. A wrap-around shield protects you from burns when handling the smoker, and provides a convenient place to store the hive tool.

Veil

A veil will protect the face, and especially the eyes, from stings. It should be made of gauze stiff enough to stand away from the face and protect it from stings. The front panel of gauze must be black or dark-coloured, otherwise it is very difficult to see through.

Single-piece suits that incorporate veil, hat and overalls are very popular. They are best for hobbyist beekeepers as they are more bee-tight than the separate units, provided all the zips are closed and the zip joints are covered properly. The veil can be unzipped when not needed. You can also get a half suit that incorporates a hat, veil and jacket.

Some beekeepers like to wear a base-ball cap under their bee suits, to keep the veil off the face and prevent bees stinging

Fig. 2.1 Having a full set of protective clothing will make your beekeeping much more enjoyable.

the nose and forehead, especially on windy days when the veil can be pressed against the face.

Some beekeepers prefer to use separate veils, hat and overalls, and these can be cheaper especially if you already have a hat with a rigid brim. A hat worn with a veil should have a stiff brim to keep the veil off the face, and should be light and comfortable to wear. Don't use a felt hat, as bees will become entangled in the fabric and start stinging.

The best veil to buy is the round model, as it is easier to store when not in use. Square veils should be folded flat when not in use, to help keep their shape. Some bought veils have very long drawstrings. This method of fastening is inconvenient, and it is better to either buy veils with elastic ties or replace the drawstrings with elastic.

Veils will give more protection if they are put on before the overalls, so the bottom is tucked in, though this is less convenient if you are going from apiary to apiary.

Overalls

If you don't use a one-piece suit, a pair of overalls is a convenient way of excluding bees and keeping your clothes clean. Overalls should be light in colour for coolness and have a zip front. Sew up the side openings and make elasticised cuffs at the wrists and ankles, to prevent bees from getting inside.

Blue overalls (particularly dark blue) should not be used, as this colour is very conspicuous to bees. Aggressive bees may sting someone wearing blue more than they sting others nearby.

Boots

If you wear ordinary ankle boots, use gaiters to protect your ankles from stings, or gather the surplus cloth in each overall cuff and tie it into the knotted boot laces. Gumboots can be made perfectly bee-tight if you put the trouser leg inside the boot and the overall leg on the outside, and use an elasticised cuff or rubber band around the bottom of the overalls.

Gloves

We recommend that new beekeepers wear gloves. They can be dispensed with as confidence grows, provided conditions are satisfactory — though many experienced beekeepers often wear gloves when they are in a hurry, or when the bees are aggressive. A few beekeepers develop allergies to propolis, and find they have to wear gloves as a result.

There are few adequate substitutes for leather beekeeping gloves. These are made of smooth leather and have long, elasticised cuffs to protect the wrists. Investment in a good pair of gloves is worthwhile, since they will last for a long time if treated well.

Rubber gloves are satisfactory for short periods but will make your hands sweat. If you do use them you should wear cotton gloves inside the rubber ones and wash them after each use. If you wear plain rubber gloves without the cotton gloves these should be sanitised after each use to avoid the risk of infection. Surgical gloves also work, but are not very robust.

If you don't wear gloves, stop bees from walking inside your sleeves by rolling up your sleeves or using elasticised cuffs or separate armlets.

Hive tool

This toughened steel lever has a flat, thin blade used for prising apart hive components and scraping them clean. Several models are available. The most commonly used is probably the Kelly, which has a right-angled bend in the blade at one end. The Maxant type has a hook at one end, which is useful for lifting up frames.

Some people make their own hive tools, or use equivalent tools purchased from hardware stores. The most important part of a good hive tool is its thin blade, so it can easily be inserted between boxes. It is a false economy to use an inadequate tool such as a screwdriver, which will damage the boxes and make working hives more difficult.

ACQUIRING BEES

There are several ways of obtaining your first bees. You can purchase a nucleus colony, buy an established hive, capture a swarm, or even transfer a colony of feral (wild) bees into a hive.

If you buy a hive or nucleus colony, make sure the vendor is registered with the American foulbrood pest management strategy management agency or its contractor AsureQuality. This is a legal requirement to prevent the spread of diseased bees or beekeeping equipment. Beekeeper registration does not guarantee that the hive is healthy, so the hive should still be thoroughly checked before purchase.

Nucleus colonies

One of the best ways to start beekeeping is to purchase a nucleus colony (called a nucleus or 'nuc'). A nucleus is a small colony of bees including a queen bee, which is housed on several standard frames (often four). Nucleus colonies have the advantage that only a small unit has to be dealt with at first, and your confidence can grow as the colony expands.

It is best to start with at least two nucleus colonies as an insurance against one queen failing. If this happens you can unite the hives or swap brood between them.

Nucs may be obtained from a local bee-keeper or from a queen bee producer. A nucleus colony usually comes without a hive, and must be transferred to your own equipment from the wooden or cardboard nuc box it came in. This is easy to do provided the following basic steps are taken.

Fig. 2.2 Transferring a nucleus to a full-sized box.

When the nuc box arrives, put it in a cool, airy, dark place for an hour or so to let the bees settle down. Meanwhile place a prepared hive, with a feeder and enough frames to fill the box when the nucleus is installed, on its permanent stand.

The best time to hive a nucleus is late afternoon or early evening, so the bees will settle down without much flying. Carry the nucleus to the permanent hive. Take the frames out of the prepared hive. Puff smoke gently over the bees in the nuc to quieten them. Then lift the frames out of the nucleus one at a time, with the bees attached, and place them together next to the feeder in the new hive. Brush or shake any bees remaining in the nucleus box at the entrance of the hive, and place the remaining frames of foundation or combs in the hive. Half-fill the feeder with thick sugar syrup (two parts by weight of sugar to one part of water), and replace the hive mat and lid.

When all the bees are inside the hive, place a block of wood across the front to reduce the width of the entrance to about 50–75 mm. Keep this block in place until the colony has increased in population and occupied more combs in the brood chamber. Depending on the strength of the colony and the seasonal conditions, the entrance may be gradually widened to its full size.

If there is any danger of robbing at the time you introduce your nucleus colony, once most of the bees are inside your hive place green grass *loosely* in the entrance. The grass will deter robbers, and it will wilt in a day or so and slowly release the bees. Alternatively, you can return the next day and remove the grass.

Continue feeding with the sugar and water solution at weekly intervals if the weather is cool or wet, but feeding can be stopped as soon as the bees are able to gather enough nectar from natural sources.

Don't feed more syrup than the bees are able to take up and store overnight. Any surplus in the feeder may attract robber bees, with fatal results to the nucleus colony. Feed in the evening and be very careful not to spill any syrup around the hive. When the bees are well established and there is a continuous nectar flow, the feeder can be shifted

Fig. 2.3 Buying established hives is the fastest way to start beekeeping.

to one side of the brood box, and later removed and reinstalled in the second brood box when this is added. Frame feeders should be replaced by a frame of drawn comb if possible, or comb foundation if no drawn comb is available.

Once the bees occupy all the frames, you should add a second brood box.

Established hives

Buying established hives is a simple method of starting beekeeping, and is a faster way to get surplus honey. While there is not a steady trade in beehive sales, advertising in a local newspaper, beekeeping journal or online trading site may help you to find a vendor.

As with nucs, it is best to start with at least two hives as an insurance against one queen failing. If this happens you can unite the colonies or swap brood between them. Having more than one hive also enables comparisons to be made, but it's prudent to not keep more than four or five hives until you have gained a full season's experience.

A beehive in good condition should have:
- plenty of bees covering most of the combs in the brood boxes (usually two)
- plenty of brood (except in winter), and honey and pollen stores
- no American foulbrood disease (AFB)
- good-quality combs (mostly worker comb)
- sound woodware of standard dimensions.

Go through the hive and examine it carefully, preferably with an experienced beekeeper.

Swarms

Honey bee colonies reproduce and disperse naturally by swarming. Particularly in spring, swarms containing thousands of worker bees and a queen (usually the colony's old one) leave hives looking for new nesting sites. There is another type of swarm. Absconding swarms in the summer and autumn — when much of a colony's population leave the hive — are a new phenomenon now that varroa is established in New Zealand. These swarms are a liability as they are usually infested with varroa, are often small in size and will struggle to survive over winter.

Capturing swarms is a cheap and usually simple way of starting new colonies, although it does have drawbacks. Swarms may carry varroa and spores of American foulbrood, and you cannot tell this by simply looking at a swarm. Because of the risk of American foulbrood, swarms are best hived in old hive parts, and located in a separate apiary away from other hives for the first season. A further drawback is that swarms usually contain a colony's old queen, and so will need to be requeened during the first summer. They should be treated for varroa as well.

Late-summer swarms are difficult to overwinter. Any swarm after January that is smaller than a soccer ball will rarely survive the winter unless united with another colony.

Capturing a swarm

If someone tells you about the location of a swarm, check with the property occupier before setting out to ensure it is still there. Swarms may cluster in a spot for only a few hours, or they may stay for days.

The most co-operative swarms will settle on a low branch of an accessible bush or tree. Capturing them can be simple, but remember it is not true that swarming bees never sting. Swarms are usually gentle but some can be aggressive, particularly if they have been clustering for several days.

Place a large cardboard box under the swarm and as close to it as possible, and give the branch a sharp knock. As soon as the bees have fallen into the box, replace the lid or put a sack over the top. The bees

Fig. 2.4 A swarm ready for collection.

may then be taken to their new location and left in a shady spot until evening, when they can be hived. If, on the other hand, a lot of bees are flying around after the swarm has been knocked into the box, leave the box partly open on the ground near the clustering site and wait until evening to move it.

Instead of a cardboard box, you can use a hive box, a plastic bucket with a ventilated lid or even a sack made of woven synthetic material as a receptacle for the swarm. Don't use a plastic bag as the bees will suffocate. Buckets are handy if you are working up a ladder.

Fig. 2.5 A swarm enters its new home in a beehive.

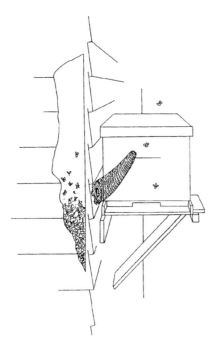

Fig. 2.6 Using a bait colony to capture bees from a feral colony established in a cavity.

If your swarm is clustered in an inconvenient place, such as on a post or on the walls or eaves of a building, hold the container close to the swarm and beneath it. Use a bee brush or a flat piece of wood to scrape the bees into the container. Partially cover the container and put it on the ground nearby. Any bees still on the clustering site should be brushed onto the ground near the container. If the flying bees enter the box, the queen is inside too. The box should be left until evening, and then moved to its new location.

If the queen is not in the container, most of the flying bees will return to the clustering site. In rare cases half the bees will stay in the container, and the other half will return to the clustering site — this usually indicates that more than one queen is in the swarm. If the bees behave in either way, the whole capturing process must be repeated until it is successful.

Hiving a swarm

The best time to hive a swarm is late afternoon or early evening, as the bees are less likely to leave the hive or annoy neighbours with orientation flights.

Prepare a single-storey hive with eight or nine frames and a frame feeder, and put it in the new hive's permanent location. If you are using a top feeder, fill the box with up to 10 frames. Use old combs if available, because of the risk of American foulbrood — a colony that later exhibits symptoms of this disease will have to be burnt, along with the hive.

Place a wooden ramp or a sack up to the landing board so the bees can enter the hive easily. Dump the container of bees on the ramp with a short, sharp jerk. The bees will very soon move into the hive in a continuous stream, in what must be one of the most spectacular sights in beekeeping. Sometimes the queen can be spotted amongst the workers as she marches in.

If there is no nectar to be gathered, you will need to feed the swarm. A large swarm (say twice the size of a soccer ball) that is hived at the beginning of the main honey flow will usually store a surplus of honey.

Feral or wild colonies

Hives may also be started from feral (wild) colonies. The job of transferring bees and combs from somewhere like a wall lining to a standard beehive is difficult and messy, but some new beekeepers are prepared to do it. Feral hives can also harbour American foulbrood, so if a feral colony is not wanted for starting a new hive it should always be destroyed.

The best time to hive a feral colony is in spring, as the bee population is low and the nest contains a minimum of honey. In summer or autumn the operation can be a lot messier as the opposite is true, and robbing will often start once honey is exposed.

There are two ways of removing bees from feral colonies: either the bees only are taken, or the whole colony is salvaged.

If the whole colony is to be removed, first prepare a hive with some frames containing foundation, and some frames that have no wire or foundation. Place the hive as close to the feral colony as possible. Next open up the feral colony's nest and expose the bee combs. Before taking any further action, examine the brood carefully for signs of American foulbrood. If this disease is present, the combs and bees must be destroyed and the incident reported to the American foulbrood pest management agency or AsureQuality. If the colony is disease-free, it may be salvaged.

Honey combs can simply be shaken free of bees and put into a large plastic bag, and later eaten as comb honey or squeezed to give liquid honey. In some areas there is a danger that wild honey may contain toxins from the tutu plant, so in these areas the honey should be used inside the new hive for feed or destroyed (see Chapter 20).

Cut the brood combs to size and fit them to unwired frames. It is best to tie them in place with string made from natural fibre, as this will be removed by the bees as they fix the combs in place with wax. Put the frames containing brood comb into the hive, along with as many bees as can be scraped or brushed in. Fill up the rest of the box with frames of foundation and replace the lid.

The cavity formerly occupied by the bees should be scraped clean of wax and honey, scrubbed with water and a strong-smelling disinfectant, then dusted with an insecticide powder and thoroughly sealed. This is necessary to stop another swarm being attracted into the same cavity by the smell of beeswax and honey. Expanding foam aerosols, fibreglass ceiling insulation or sheep's wool can be used to block large entrances to cavities.

When the bees have clustered in their new hive that night, remove them at least 5 km away to stop the flying bees returning to their previous home. Requeen the colony in the first season after transfer, and cull the old combs after new ones have been filled.

A simpler method is to remove the flying bees only, and use them to boost the strength of an existing small colony. Place a one-way trap over the nest entrance, such as a gauze cone about 250–300 mm long with an entrance just large enough for a bee to crawl through (Fig. 2.6). Put a bait colony very close by. Foraging bees can leave the feral colony but cannot re-enter through the trap, so over a period of several weeks they will join the bait colony and the feral colony will dwindle away.

Though an easier method, this technique has drawbacks:
- no check for American foulbrood is possible
- it is successful only if all alternative entrances can be blocked
- unless the nest cavity is later cleaned out or thoroughly closed off, it is likely to be used in the future by another swarm
- remaining honey may ferment, creating off-odours and sometimes a sticky mess.

If you are planning to remove a feral colony from somebody else's property, be sure to have the permission of both the property occupier and the owner if they are different people. Before you begin work have a clear understanding of whose responsibility it is to repair any damage done.

Package bees

A package of bees is like a nucleus colony without the frames, and consists of a quantity of bees (often 1–1.5 kg) and a queen. Packages are not very common in New Zealand, but are included here in case beekeepers can source them.

Packages are similar to nucleus colonies in that they must be housed in assembled hive equipment, and they should be obtained in spring to allow enough time to build up in strength before winter. They are much less effective than nucleus colonies though, because they contain no brood so take much longer to build up in size.

When a package arrives, place it in a cool, dark place for an hour or so until the bees settle down. As with a nucleus colony, you will need to have prepared a hive on its new site. It should have an entrance reducer on and be fitted with a feeder.

Choose late afternoon or early evening for hiving a package. Remove three frames from the middle of the hive. Take the queen cage and feed can out of the package, and quickly dump the bees into the hive. Bees adhering to the inside of the mesh can be dislodged by banging the solid corner of the package container on the ground, and then pouring the dislodged bees out the feeder hole. Once the bees are in the hive, carefully replace the three frames.

Expose the white candy in the queen cage by removing the plastic cover and wedge the cage between the two middle frames in the hive, at about the centre of the frames. The exit hole in the candy end should be slightly uphill, so if any attendants die after the cage is introduced they don't roll down and block it. If there is mesh on one side of the cage only, put that face-downward so that any honey running down the comb doesn't enter the cage and cover the queen. Replace the hive lid. Leave the empty package lying just outside the entrance overnight, to allow stray bees to enter the hive.

After a week make a very brief visit to see if the queen is laying. Replenish the feeder, but otherwise don't disturb the colony. Make a second visit after another week to check there is adequate food.

LOCATING HIVES

A well-sited apiary is one that suits the bees and the beekeeper, and doesn't inconvenience neighbours or passers-by. Where apiaries are located, and how far apart they must be, are questions that should be sorted out by beekeepers and landowners together. In general, apiaries of more than about five hives are located at least 1.5 km apart.

In some built-up areas beekeeping is subject to council bylaws. If you intend to keep bees in an urban area, first find out if your local council imposes any restrictions. Some councils require you to register your apiary. They may have the power to decide if your location is suitable for beekeeping, and may charge an annual fee for this. They may also require you to get the written permission of all your neighbours first. Local bodies without specific beekeeping bylaws still have power to act against beekeepers if necessary, under general nuisance provisions in relevant legislation.

If there are no local bylaws regulating beekeeping, good hive management and public relations will help this fortunate state of affairs to continue.

Honey produced in the North Island and northern regions of the South Island and offered for sale (including barter) or export is subject to regulations designed to minimise the risk of the toxin tutin being included in the honey. See Chapter 20 for details on what you must do if you keep hives in these areas.

SELECTING THE APIARY SITE

Choosing where to site an apiary is one of the beekeeper's most important tasks, particularly as the location can have far-reaching effects on the honey crop produced. Too often honey bee colonies are simply placed close to a particular nectar source, and little thought is given to other factors.

Permission

It is essential to obtain the permission of both the land owner and occupier.

Apiary sites on farms should be arranged in full consultation with the farmer. Discussions should include any plans for future use of the land, as it is difficult to gain access to hives across paddocks that have been shut up for hay or cultivated for a crop. Farmers will also know of flood-prone spots. Ensure that farmers have your name, address and phone numbers, so they can contact you if hives are damaged by stock, storms or vandals. Farmers will also want to warn you of hazards on their property.

Special permits may be required to site hives on land administered by forestry companies or by government agencies. Sometimes beekeeping rights on forestry

Fig. 2.7 A sheltered domestic apiary.

land are tendered for, but purchasing hives on such property does not necessarily guarantee continued rights to the site.

Shelter

Protection from prevailing winds is essential. Hives should be located in the lee of patches of bush, shelter belts, stop banks or whatever shelter can be found in the area. If there is none, either put up artificial shelter or plant some quick-growing shelter trees. You must also be prepared to trim hedges yourself if your hives prevent access by contractors.

Although apiary sites should be sheltered, they should not be at the bottom of deep gullies where access can be difficult and cold, damp air will lie. Some air drainage is essential, to prevent dampness in hives and rotting of hive woodware.

Sunlight

Apiaries should receive as much sunlight as possible, especially in the morning as hives that don't receive sun until the middle of the day miss out on a lot of potential foraging time. Take particular care to ensure that sites are not shaded in winter when the sun is at its lowest. Hive entrances usually face north, but this is not critical.

Food sources

For bees to produce surplus honey the apiary must be within flying range of good nectar and pollen sources. Bees will fly several kilometres from the hive, although foraging is more efficient if hives are close to food sources. The availability of spring forage is also important, as early nectar sources will result in better colony development and will reduce the need for costly sugar feeding.

Fig. 2.8 Take care to locate your apiary where it won't be flooded.

Pollen substitutes are available but can be expensive to buy and feed. Adequate sources of pollen near an apiary are vital.

Access

A common mistake among domestic beekeepers is to locate hives without giving thought to access. Hives on garage roofs or on steep banks may look good, but harvesting full honey boxes from such positions is difficult, and often dangerous.

Make sure you can at least take a wheelbarrow to all your apiaries, and that out-apiaries have vehicle access for most of the year. It could be costly if you are cut off from a site for several weeks at a critical time.

To reduce the risk of vandalism or theft it is sensible to conceal apiaries from public roads. This can usually be achieved without going to the most remote spot on the farm.

Some new beekeepers with access to a lot of land often feel they need to spread their hives singly or in small groups all over the property. This makes access and servicing the hives quite difficult and slow. Remember that bees fly over gates and rivers more quickly and easily than you can drive through them, so site out-apiaries as conveniently as possible and put the bees in only one or two locations.

Arranging hives

The way hives are set out on a site affects how easily you can manage them, and how easily the bees can find their particular hive. Good layout of hives within an apiary can make for much more efficient management. The layout chosen will be determined by factors such as the size and shape of the site, the type of vehicle used, and whether the site has to be fenced to prevent stock damage.

Hives are best placed in twos or fours, with a good 1.5 to two metres to the next group of two or four. If stock have access to the apiary, groups of two, and especially four, hives can offer some support to each other. A suitable space between the groups means you can work from the side of the hives and have room to put supers beside them. Hives jammed together are very difficult to work, and hard on the back as well.

The number of hives in an apiary is influenced by the nectar and pollen sources in the area, and how many colonies these food sources can support in the poorest part of the season. This is hard to determine though, and in practice the number of hives in a site also depends on the type of vehicle a beekeeper uses, the number of people working the hives and individual preference.

Most beekeepers find 15–25 hives enough on one site — by the time these have been inspected, the bees may have begun robbing, and the beekeeper will need a rest. Commercial beekeepers' crews will work apiaries of 30–35 hives, and in South Island beech honeydew areas, apiaries can contain hundreds of hives.

You also need to place hives so bees can easily find them. Traditionally beekeepers placed hives in straight rows. This layout looks neat and is easy to work, but it confuses the bees. They become disorientated and return to the wrong hives — a process called drifting.

Drifting follows certain patterns. If hives are placed in straight rows, bees drift to the ends of rows. Where two or more rows are used, the front row collects bees at the expense of hives in the rear.

Heavy drifting means that some hives may become so depleted of worker bees that they gather little surplus honey, while others become overcrowded and may swarm. Apiary management is made more difficult, as each hive must be treated individually rather than all hives in an apiary dealt with in a similar way. Selection of breeding stock based on honey production records is not reliable in apiaries where there is a lot of drifting. Drifting bees will also fight when there is no nectar flow.

You can minimise drifting by:
- arranging hives in irregular patterns, or at least with entrances facing different directions (Fig. 2.9)
- painting hive boxes different colours
- leaving some landmarks, such as bushes, in an apiary.

Fig. 2.9 A commercial apiary set out in a way that minimises the drifting of bees between hives.

REGISTERING AN APIARY

Every place where bees are kept is called an apiary, and all apiaries must be registered with the management agency for the American foulbrood pest management strategy, or its contractor AsureQuality. An apiary must be registered if beehives have been placed there for more than 30 consecutive days, even if only one hive is involved. Registration assists with the management agency's programme to control American foulbrood, and also with MAF Biosecurity New Zealand's programmes to prevent the establishment of exotic pests and diseases.

All apiaries must be identified with the code number allocated when the beekeeper first registers, which can be placed either on a hive or on a sign in the apiary. Code numbers help in the identification of the apiary owner.

See Chapter 20 for more information about apiary registration.

GOOD PUBLIC RELATIONS

Please consider others when you keep bees, and help to give beekeeping a good name. Don't put hives close to roads where bees may annoy pedestrians or passing motorists. On farms take every care not to interfere with farmers' activities — it is after all their property.

The main complaints about beekeeping in urban areas are:
- bee droppings soiling houses, cars and laundry (especially in spring)
- swarms settling on private property
- people being stung or buzzed by bees when hives are disturbed, or when robbing starts
- bee flight paths through neighbouring properties
- bees collecting water from taps, swimming pools or washing.

Almost always the beekeeper is at fault. There are a dozen basic rules for good beekeeping in urban areas:
- Keep no more than two or three hives on a residential section.
- Position your hives so they don't become troublesome. They should be in a

sunny, sheltered spot that cannot be seen by neighbours — most people don't worry about what they can't see. Don't place hives close to your neighbour's house or driveway, or near frequently used areas such as vegetable gardens or clotheslines.

- Force bees to fly at least 2 m high as soon as they leave their hives, by placing the hive entrance within a few metres of a screen such as a fence, trellis or hedge. This keeps the bees flying above human head height and so helps to prevent 'buzzing'.
- If the bees do establish flight patterns that are a nuisance, you will have to act at once. One option is to remove the hives at least 5 km for three to four weeks or so before returning them to the original site, so that the foraging bees are new ones unfamiliar to the area. The period of exile will need to be longer in winter. Another option is to leave the hives on-site but rotate them through 180° one night, and block the entrances loosely with grass. This may force the establishment of new 'bee lines'. Hives can also be relocated on a section by moving them all together, but no more than 1–2 m each day.
- Provide water within several metres of the hive if no natural sources are available. This will reduce visits by your thirsty bees to neighbours' swimming pools, wet washing and dripping taps. Provide water by letting a tap drip very slowly on to sandy soil, or leaving a container full of regularly-changed water. Open water must have suitable floats (e.g. wood shavings, polystyrene blocks, or an aquarium plant) for the bees to land on.
- Keep a gentle strain of bee in every hive. Pure strains of yellow or leather-coloured Italian bees are the best in New Zealand conditions. The Carniolan strain can also be very gentle, although since the bees are black in colour they may appear more alarming to the public than the Italian strain. Carniolans also swarm more often. All queen bee producers supply gentle strains of bees for hobbyist beekeepers. Requeen temperamental hives promptly, especially ones that have requeened themselves, and requeen all hives every one or two years.
- Work bees only during the warmest part of the day. Bees are quietest during a nectar flow and a period of fine, warm weather.
- Once you can handle your bees with confidence, try not to wear gloves while working hives in towns or cities. Stings on the hands are easily removed and the pain quickly passes. Stings on gloves, however, are not felt and the scent associated with the sting encourages other bees to sting. You won't feel the stings, but neighbours might. Perhaps more importantly, working barehanded teaches you how to use the smoker properly.
- Practise proper swarm prevention techniques.
- Don't allow bees to start robbing exposed honey or sugar syrup. Feed syrup and put wet (sticky) honey boxes on hives only in the late evening.

- Be a good beekeeping neighbour. Collect any swarms quickly, after first advising others (especially children) not to interfere with the swarm.
- Stress the value of bees in pollinating fruit trees in the neighbourhood, and share some of your honey crop with those living nearby.

Try to begin your beekeeping on the right footing, with quality protective clothing and bees that will enable your hobby to get off to a good start. Select apiary sites that give bees easy access to nectar and pollen, and that allow you to manage the hives easily and without inconvenience to others. Time spent getting things right now will save inconvenience or beekeeping failure later on.

Hive equipment  3

Beekeepers will need to develop at least rudimentary woodworking skills. Though some hive equipment can be purchased already assembled, most beekeepers assemble kitset hive parts, and many make at least some hive components themselves. Some beekeepers end up as very competent carpenters by making hive equipment.

Hive equipment is expensive, so correct assembly and preservation — and ongoing maintenance — are important to protect your asset.

HISTORY

Honey bees have been kept in many different types of hive in different parts of the world. The first bee colonies in New Zealand arrived in straw skeps, the traditional European beekeeping container with the classical 'beehive' shape. During the 19th century, bees in New Zealand were kept in skeps or simply in wooden boxes — kerosene or benzene cases were particularly popular. All these early hives were 'fixed-comb hives', in which bees were allowed to build comb attached to the hive interior. The bees were usually destroyed to harvest the honey, and the hives repopulated by catching swarms.

In 1851 an American clergyman, L. L. Langstroth, designed an improved hive based on an important principle now known as the 'bee space'. Langstroth's hive contained frames hanging within a box, and surrounded on all sides by a gap of 6–9 mm (Fig. 3.1). That gap, called the bee space, is left open by the bees and used as a passageway. Larger spaces are usually filled with wax comb, while smaller spaces are filled with propolis.

Langstroth's innovation permitted full manipulation of the colony and the harvesting of honey without killing bees. His basic design is used in all major beekeeping countries today.

The first movable-frame hives were brought to New Zealand in 1876, and more were made here within the

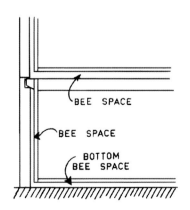

Fig. 3.1 The arrangement of frames in a hive to maintain bee space.

next few years. The Langstroth hive was advocated as a standard hive for New Zealand in 1881. Early adoption of Langstroth's design as the only movable-frame hive has meant that New Zealand is free of the large numbers of different designs that exist in some other countries. To ensure this advantage continues, it is strongly recommended that beekeepers use only hives that are built to standard specifications, which are listed in Table 3.1. Following the standard also means that all hive parts are interchangeable among hives.

Table 3.1 Standard dimensions for the New Zealand Langstroth hive (millimetres)

Hive body				
Timber thickness	20 (derived from 25 mm green-sawn)			
Outer dimensions	505 x 405			
Inner dimensions	465 x 365			
Depth				
Full depth	240 (derived from 250 x 25 green-sawn)			
Three-quarter depth	185 (derived from 200 x 25 green-sawn)			
Half depth	133 (derived from 150 x 25 green-sawn)			
Rebate depth: 13				
ledge: 10				

Frames	Hoffman (full depth)	Hoffman (¾ depth)	Manley (¾ depth)	Section holder
Top bar				
length	482	482	482	482
width	25	25	25	38
thickness	16	16	16	8
lug thickness (at point of contact)	10	10	10	8
Bottom bar				
length	450	450	450	450
width	25	25	25	38
thickness	10	10	10	8
End bar				
length	230	175	175	125
width	33	33	43	46
thickness	10	10	10	8

Section size				
overall length	428.6			
width	46			
thickness	3			
folded size	108 square			

SOURCES

Hive equipment can be purchased in kitset form or made in home workshops. If you are going to make hive equipment yourself, remember that frames and boxes, in particular, must be constructed with millimetric precision to ensure correct bee spacing, especially in their internal measurements.

The hive dimensions in Table 3.1 were settled on when New Zealand adopted the metric system in the early 1970s, and the standard 1" thick timber (dressed to 7/8" or 22 mm) was changed to 25 mm thick timber (dressed to 20 mm). The internal hive body dimensions are critical to preserve the bee space around the frames. Some beekeepers and equipment manufacturers prefer to stay with 22 mm thick timber, in which case the internal dimensions should remain the same but the outer dimensions will become a few millimetres larger (509 x 409 mm).

TYPES OF TIMBER FOR HIVE CONSTRUCTION

Radiata pine (*Pinus radiata*) is the timber most commonly used for beehives in New Zealand. It is readily available, light, and easy to work with. Tanalised timber (including marine plywood) should not be used for hive construction, as arsenic leaching out of the wood is toxic to bees. Tanalised timber can be used only as ground runners on floorboards, or for the outside parts of telescopic lids, where bees do not have any contact with the timber.

Virtually any other kind of untreated timber, native or exotic, is suitable for hive construction, including finger-jointed planks. Caution is needed with panels of composite material such as medium-density fibreboard (MDF, e.g. Customwood), as these usually contain adhesives in the fibreboard that are toxic to honey bees. In addition, the boards do not weather well and soon fall apart.

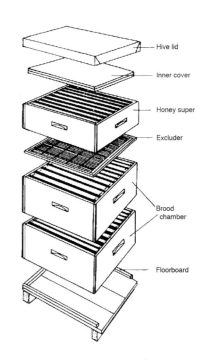

Fig. 3.2 The components of a beehive.

CHOOSING THE RIGHT HIVE EQUIPMENT

A beehive consists of a floorboard, a roof, and a variable number of boxes or hive bodies in between (Fig. 3.2). The number of boxes in the hive depends on the strength of the colony in it and the time of year.

Boxes

The bottom one or two boxes contain the colony's brood nest or brood chamber, and are called brood boxes. There young bees are reared and honey and pollen are stored for winter. The boxes above the brood chamber are for the storage of surplus honey to be harvested by the beekeeper. These are termed honey boxes or 'supers', as they go on top of (or are superimposed on) the brood nest.

Traditionally New Zealand beekeepers have used 'full-depth' bee boxes, 240 mm deep. Concern about lifting heavy weights, and the greater availability of smaller timber sizes, have led some beekeepers to introduce 'three-quarter depth' boxes, which are 185 mm deep. Some beekeepers still use these for honey supers only, but honey bee colonies will live just as well in three-quarter depth brood boxes as in full-depth. Full-depth boxes that are beginning to rot in the bottom corners can be cut down and recycled as three-quarter boxes.

Amateur beekeepers should consider using hives composed entirely of three-quarter depth boxes. They are completely interchangeable, lighter in weight, and are easy to manipulate. These advantages can outweigh possible drawbacks, such as the need for a greater number of boxes to contain a given quantity of honey, and the extra work and possible extra contractor costs to extract the honey. The move to pallets and mechanical cranes has made lifting heavy full-depth boxes easier for commercial beekeepers, but bad backs are still an industry problem.

There are two other less common types of bee box. The half-depth box, 133 mm deep, is used for producing section comb honey, and by those beekeepers unable to lift even a three-quarter depth box. A few commercial beekeepers producing cut comb honey may use a special super, usually 150 mm deep.

Boxes used to be made either with shallow (13 mm) rebates or deep (20 mm) rebates for the frame lugs to hang on. The deep-rebate boxes were designed to be fitted with a metal runner, so the frames finish up at the same height as in shallow-rebate boxes (Fig. 3.3). The purpose of the metal runner is to reduce the contact area that can be propolised by bees, so making the frames easier to manipulate. In practice, however, shallow-rebate boxes are just as easy to work and are less complicated to make. The space behind a metal runner also acts as a water reservoir, which may cause the box rims to rot. We suggest you use only shallow-rebate boxes without metal runners.

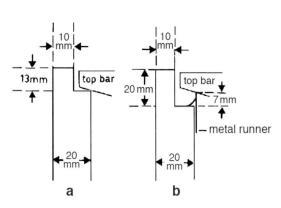

Fig. 3.3 Rebate dimensions:
(a) without metal runner, (b) with runner.

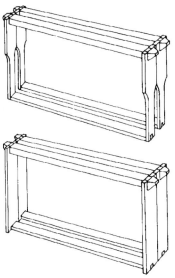

Fig. 3.4 Hoffman (top) and Manley self-spacing frames.

Wooden frames

Two main styles of frames, the Hoffman and the Manley (Fig. 3.4), can be bought from bee equipment stockists. Hoffman frames are the most common as they are used in all brood boxes and in honey supers. The end bars are scalloped to reduce the area of frames that can be propolised together, and to allow the bees to move around the ends of the frames. In the brood nest, Hoffman frames should be pushed together to give the correct comb spacing and maintain the bee space. Hoffman frames come in full, three-quarter or half-depth sizes.

Hoffman frames with 33 mm end bars should be used, as this size is closest to the bee's natural comb spacing. These frames can be used comfortably at 10 per brood box, even after a season or two of use when bees have added propolis to the end bars (Fig. 3.5).

Beehive manufacturers still offer Hoffman end bars that are 35 mm wide, a hangover from the days of imperial measurements. Ten of those frames will fit into a box when new, but not after a couple of seasons of use. This means that in practice brood boxes generally contain only nine frames, the outside ones having very fat combs (Fig. 3.6). Frames with 33 mm end bars overcome these disadvantages to brood nest management.

Whether you use 33 mm or 35 mm frames, always keep all the frames squeezed together and space the outside two frames to best fill any gap between the frames and the hive walls. Hoffman frames in honey supers are spaced evenly at eight or nine to the box, to ensure the bees extend the comb beyond the frames. This makes uncapping honey frames much easier.

Fig. 3.5 Ten Hoffman frames with 33 mm end bars, self-spacing in a brood chamber.

Fig. 3.6 Nine Hoffman frames with 35 mm end bars: the seven central frames are pushed together and the outside two evenly spaced.

Most wooden frames come with four holes in the end bars for the frame wires, and we recommend using this type. Some frames have three holes, but three wires don't support the wax foundation enough, especially if you are extracting manuka honey. Frames can be purchased assembled, wired and even with wax foundation sheets inserted.

The Manley frame is a very robust frame used only in honey supers. Its advantage is that it is self-spacing at eight frames per box, so the bees make wide combs that are easily uncapped and no time is lost in the apiary spacing frames evenly in the honey supers.

However, the Manley frame is not interchangeable with brood frames, and will not fit in some extractors.

The cut-comb frame is made with straight end bars, like a Manley frame. The end bars are 35 mm wide, and 10 frames are used per super. Propolis and wax don't build up enough to widen the end bars, as the frames are cleaned each year. The depth of the cut comb frame is determined by the size of the plastic box the honey will be packed into. These frames usually have only one wire.

If you buy second-hand hives you may encounter the Simplicity frame, which has a narrow end bar — the same width as the top and bottom bars. This frame is not self-spacing, and the frames in a box must all be spaced by hand to preserve the bee space between them. They can be made self-spacing to some extent by putting U-staples into the shoulders, but this isn't very exact. These staples damage uncapping knives when extracting honey. We suggest that you replace Simplicity frames with self-spacing frames.

Plastic frames

Plastic frames in full-depth and three-quarter-depth sizes are available from several stockists, and come either waxed or unwaxed. These frames are being rapidly adopted by both hobby and commercial beekeepers. Plastic frames are good for beekeepers who produce the very thick manuka honey, and for those who don't have the time or desire to assemble and wire wooden frames and embed wax foundation sheets.

Plastic frames are more expensive than wooden ones but last for a long time under normal conditions. Plastic frames are less attractive to the bees, and should always be placed in the top brood box or honey super for the bees to 'draw out' or build new comb on. The hive should also be fed sugar syrup or be on a nectar flow for best results, as bees consume sugar to produce wax.

Plastic frames are more flexible than wooden ones and may take a bit of getting used to when working a hive. The top bar is much thinner than with wooden frames, which increases the area of comb. As a result there is more comb area available for the queen to lay in. Plastic combs slide more readily on the hive rebates than do propolised wooden ones.

If the comb becomes damaged it can be scraped off and the bees will often build good comb to replace the old. The foundation part of the frame may need to be recoated with good-quality beeswax. Plastic frames are reasonably heat-stable, but cannot withstand boiling in water. These frames cannot be sterilised if the hive is infected with American foulbrood, and must be destroyed. Burning plastic frames creates a lot of smoke and is best done at night.

Foundation

Wooden frames are fitted with sheets of comb foundation, which are sheets of beeswax embossed with the worker cell pattern. Bees add extra wax and make the cells to complete the comb, a process called 'drawing out' the foundation. Drawn-out combs don't have to be replaced each season, but are used until they become damaged or otherwise unsuitable (such as when large areas of cells become clogged with old pollen that the bees don't remove).

Two grades of foundation are commonly used by beekeepers. 'Medium brood' is used for all brood frames and all extracting honey combs. 'Thin super' grade is used for comb honey only (sections or cut comb). Other grades of foundation are available, such as light brood and extra heavy brood. The latter is used by some commercial beekeepers in areas where heavy-bodied honeys are produced, or they may even use the thicker 'manuka special'.

Plastic frames include comb foundation as part of the frame. If they are not purchased pre-waxed, the foundation needs to be coated with a thin film of wax by the beekeeper. Beekeepers apply molten wax to plastic combs using a large paint brush or paint roller. Wax can be heated in an old electric frying pan with a little bit of water added. Take care not to overheat the wax. It is best not to melt wax over an open flame, though you can use a wax container placed inside a bigger one that holds water.

Plastic foundation sheets for fitting into wooden frames are also available from suppliers, and again these come either plain or already covered with a thin coating of beeswax. The plain ones can be waxed as described above.

Plastic foundation for fitting into wooden frames is available in two different lengths, depending on whether they are to be used in frames with or without grooved end bars. Plastic foundation sheets really need a split or grooved bottom bar and grooved end bars to support them. Some beekeepers using frames without grooved end bars insert split pins or frame staples through the holes in the end bars to go on either side of the plastic foundation. Plastic frames and foundation sheets do not have or need wires.

Hive lids

Hive lids or roofs come in three basic styles. The most common is the telescopic lid (Fig. 3.7), which is a wooden frame supporting a metal cover that hangs down or 'telescopes' over the sides of the hive body. Telescopic lids are commonly used with a hardboard inner cover or flexible hive mat, to prevent the bees sticking them down too tightly to the tops of the frames. This type of lid can create problems for beekeepers who shift hives regularly, since the lids hang down over the sides of the hive body, meaning the hives cannot be fitted snugly together on a truck deck.

There is another type of telescopic lid that avoids this problem, so is much better for hives that are moved frequently. This lid is made of galvanised steel or plastic

Fig. 3.7 Telescopic (L) and flush (R) galvanised steel hive lids.

with no wooden framing (Fig. 3.7). It fits snugly over a box, so hives can be placed close together when being moved. Some versions are made from heavy gauge steel and have sprung ends, so the lid grips onto the top super. Snug-fitting telescopic lids of any type must also be used with an inner cover to prevent sticking to the top bars, but the lids do not blow off readily because of their snug fit. One disadvantage of this type of lid is that it cannot easily be used upturned for stacking boxes unless it is made of very heavy 0.95 mm gauge galvanised steel.

A third type is the migratory lid (Fig. 3.8). This has no overhanging wooden sides, but may have overhanging ends, and is held on simply by the bees' propolis and wax. For this reason these lids must not be used with inner covers or hive mats.

Fig. 3.8 Migratory hive lid.

Floorboards

Floorboards or bottom boards (Fig. 3.9) can be made with or without a landing board, and with shallow or deep risers.

Base plates for floorboards can be made of demolition timber such as tongue-and-groove flooring or sarking, plywood, dressed timber, fibrolite (fibre cement sheeting), or even galvanised steel. Tanalised timber must not be used for the base plate unless covered by galvanised or offset printing sheeting.

Floors with mesh in them (Fig. 3.10) may keep hives drier in humid climates and provide extra ventilation when shifting. Some beekeepers make ventilated floorboards by inserting wire or plastic mesh in the base of the floor, or creating a mesh rim (with an entrance) on top of an existing floorboard and reversing the floor so the bees use the entrance in the mesh rim. Ventilated floorboards can also help with varroa control, as varroa falling through the mesh cannot re-enter the hive. However, mesh floors on their own do not control varroa. Mice may damage plastic mesh, and formic acid sometimes used for varroa control

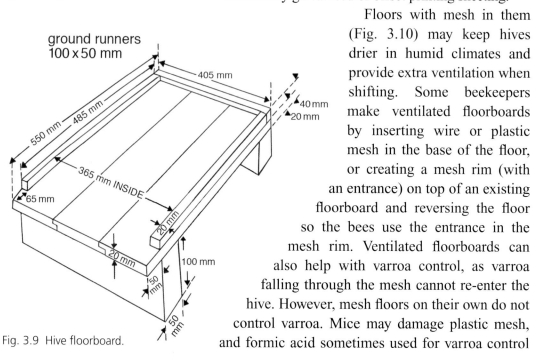

ground runners
100 x 50 mm

405 mm
485 mm
550 mm
40 mm
20 mm
365 mm INSIDE
65 mm
20 mm
20 mm
100 mm
50 mm
50 mm

Fig. 3.9 Hive floorboard.

will damage galvanised mesh. Stainless steel mesh is the best, but also the most expensive.

Landing boards can be of any length or omitted altogether. Bees quickly adapt to not having one by landing on the front of the box and walking into the entrance. Long landing boards keep grass growth away from the entrance, but make it difficult to stack hives on a vehicle if they have to be shifted.

Risers provide the entrance to a hive and can vary in depth from 12 to 20 mm. They should not be made from tanalised timber. Risers should not form a watertight fit at the back of the bottom board, so that any water pooling on the floorboard can drain away if the hive is sloping backwards.

Deep risers (more than 12 mm) provide extra ventilation during summer, but the bees will build brace comb on the bottom bars of the frames to act as ladders. If you move these frames into another box you should scrape this wax off, or it will not fit in the bee space between boxes. If a floorboard has deep risers, you should use an entrance reducer in winter, to help keep

Fig. 3.10 A screened floorboard is much more common now that varroa is here.

Fig. 3.11 Floorboard with a tunnel entrance.

out mice and make it easier for the bees to defend themselves against wasps or robber bees. An entrance reducer is a block of wood that completely covers the entrance except for a gap 8 mm high and about 100 mm wide in the centre. Shallow risers have the advantage that an entrance reducer is not needed in winter, especially if an entrance tunnel is used (Fig. 3.11).

Ground runners are essential, as bottom boards seldom last more than a few years if placed directly on the ground without them. The ground runners are usually 100 x 50 mm or 75 x 50 mm, and should be made from a ground-contact timber such as recycled totara or ground-treated tanalised pine.

Other equipment

Top mats are sometimes kept under the lid. These prevent the bees from sticking the lid down too tightly with propolis and wax onto the top super. Flexible hive mats made from hemp sacks, plastic, vinyl or even synthetic bags are common — if these are used they should be the same size as a bee box (505 x 405 mm), otherwise the mat hanging down below the lid can act as a wick in wet weather and draw moisture into the hive.

Flexible hive mats have been superseded in most cases by hardboard hive mats (also called inner covers, division boards or crown boards). Inner covers are not chewed by the bees as some flexible hive mats are, and will not act as a reservoir of moisture in the hive. They provide a bee space over the top bars of the top box, and can be used as division boards (split boards) if a notch is cut out of one end to make an entrance. Division boards are useful when requeening and carrying out other hive manipulations.

Queen excluders are stocked by bee equipment suppliers, usually complete with a wooden or metal rim. Plastic excluders are also available but as these do not have a rim, and therefore no bee space, can become clogged with wax and need frequent cleaning. Types of sugar feeders are discussed in Chapter 8, and bee escape boards in Chapter 14.

GETTING STARTED

To set up a single hive from new materials, you will need to buy (or make) the following:
- one floorboard
- two full-depth, or three three-quarter-depth, boxes for the brood chamber
- twenty full-depth, or 30 three-quarter-depth, Hoffman frames
- twenty or 30 sheets of medium brood grade foundation of appropriate size, or plastic frames
- one or two honey boxes, depending on how frequently you can extract honey
- sixteen Manley frames (three-quarter depth) or 18 Hoffman frames (if full-depth boxes are used) for the honey boxes
- sixteen or 18 sheets of medium brood foundation of appropriate size, or plastic frames
- one queen excluder (optional)
- one inner cover
- one hive lid
- one 200-gram reel of frame wire (not necessary if plastic frames are used)
- nails, tacks, wood preservative and paint as required.

ASSEMBLING HIVE EQUIPMENT

Bought hive equipment usually comes in kitset form. Hive components are fairly expensive, so to protect your investment and ensure your hives last a long time, assemble them carefully and preserve them well.

Boxes

Kitset boxes have either dovetailed (also called lock) corners or half-checked corners. Although the dovetailed or interlocking corner is stronger, it will rot out more quickly because of the large amount of end grain exposed — simple half-checked corners are better.

When assembling boxes make sure the ends are positioned correctly — with handholds outwards and both rebates up the same way. Nail corners with seven 50 or 60 mm galvanised flat-head nails per corner on full-depth boxes, and five nails per corner for three-quarter depth. Drive the nails in on a slight angle, rather than straight, to help prevent the joint from lifting when the boxes move as they absorb and lose moisture. Posidrive screws can be used instead of nails, and have the added advantage that the boxes can be disassembled at any time. They also hold the joints together much more securely.

Beekeepers rarely glue corners, but you can use waterproof glue if you wish. Two-pot marine glues are the best, as PVA woodworking glue may not be satisfactory over the long term. Fewer nails are needed on glued boxes.

Finally, smooth off the sharp edges and corners with sandpaper, a wood rasp or plane, before preserving with timber fungicide and/or paint.

Frames
Assembly

Pretty much the same method is used to assemble frames, regardless of what type they are. Various jigs can be made to speed up the job (Fig. 3.12), although most beekeepers with only a few hives assemble frames singly.

Use two 30 mm glue-coated nails to join each end bar to the top bar. Either put both nails down through the top bar into the end bar, or else place one nail vertically and the other horizontally through the face of the end bar into the end of the top bar under the lug. With the Australian

Fig. 3.12 Use of a jig to hold frames speeds up assembly.

pattern top bar, it is possible to place a smaller nail through the end bar shoulder or edge and into the side of the top bar. This cross-nailing greatly strengthens the joint, but the hole through the edge of the end bar may need to be drilled if you find you split too many frames. Smaller nails are required for this option.

The bottom bar (grooved or plain) is usually attached to each end bar with just one nail. Split bottom bars can be used with plastic foundation, and will require more nails. PVA or similar glue can be used to strengthen all joints. Electric or air-powered staple guns can also be used to assemble frames and boxes.

Wiring

To wire frames properly it is essential to have a wiring board. If you try to do the job with cramps or a vice you will probably get poor results. You should also use only the correct

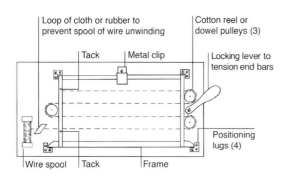

Loop of cloth or rubber to prevent spool of wire unwinding

Cotton reel or dowel pulleys (3)

Tack

Metal clip

Locking lever to tension end bars

Positioning lugs (4)

Wire spool

Tack

Frame

Fig. 3.13 Construction of a wiring board.

frame wire, which is available from bee equipment stockists.

Fig. 3.13 shows a plan for a wiring board. These boards usually cannot be bought, so either make one yourself or borrow one from another beekeeper or a beekeeping club. The base of the wiring board must be solid — a piece of 300 x 50 mm timber is good. A single wiring board can be used for both Hoffman and Manley frames, although some modifications may be necessary.

A frame is placed inside four positioning lugs on the board. The locking lever is mounted off-centre, and pushes one end bar inwards as it locks the frame. Other lever-action devices can also be used to push in the end bars. A metal or wooden clip prevents the bottom bar from rising as the wires are tensioned, and keeps the bottom bar straight. The wire is threaded through a staple or inner tube fixed to the board to stop it unravelling.

Three pulleys make it much easier to operate the board. These should not have a rim, so use either cotton reels cut in half (with the cut surface uppermost) or 40 mm lengths of 25 mm dowel drilled to take a 75 mm jolt head nail.

To wire a frame, mount the wiring board firmly on a workbench. Place a frame in the jig and partly nail two frame-wire tacks into the edge of the end bar, near the first and last holes. These tacks should not be placed directly above the holes, as they could obstruct them and make later rewiring difficult. Two staples from a hand staple-gun can also be used.

Thread the wire through all the holes, and wind the free end around the last tack. Drive the tack into the wood and twist the wire several times until it breaks. Don't cut it, or a jagged edge will later snag fingers. Alternatively, staple the wire then bend the wire back on itself and apply a second staple close to the first one and over both lengths of wire before breaking off the wire.

Tension the end bar with the locking lever. Slip the wire off the pulleys and wind the slack back onto the spool. Pull each wire taut with your fingers or a pair of pliers. Wind the wire around the second tack and drive the tack in, or staple the wire as before, and break off the wire as previously described. When the locking lever is released extra tension is put on the wires.

If wired frames are left for some time before use, the wires may loosen. They can be tightened temporarily by putting a small kink in the wires on the end bars with a pair of long-nosed pliers, or more easily by using a frame wire crimper (available from bee equipment stockists). Some beekeepers put metal eyelets in the end bar holes to prevent the wire cutting into the wood and causing the wires to become slack.

Embedding foundation

The only satisfactory way of embedding the wires into a sheet of foundation is to use an electric embedder, which can either be bought or made at home.

You will need an embedding board (Fig. 3.14) made of 20 mm timber, particle board or fibreboard small enough to allow a frame to be easily laid over it. A 12-volt current is used to melt the wax around the wires. A suitable source might be a transformer (of the kind used for low-voltage electrical appliances), a motorcycle battery, a car battery (not fully charged), a battery charger, or dry cells.

Fig. 3.14 Embedding wax foundation with low voltage electricity.

Before embedding foundation in the frame wires, remove it from storage and leave it at room temperature (20–25 °C) for about 24 hours. Cold foundation is brittle and breaks easily.

To embed a frame, first stand it upside down on the bench with the bottom bar in the air. Fit the sheet of wax into the foundation groove in the top bar. While holding the sheet of foundation in the top bar, carefully lay the frame down on the embedding board with the foundation underneath. The wires should be supporting the weight of the frame, with the end bars clear of the bench. Check that the foundation hasn't slipped out of the top grove before you embed the sheet.

You embed the sheet of foundation by passing electric current through the wires and gently pushing down on the frame. Each wire can be embedded singly, or all wires can be embedded at once by putting the probes on the two frame wire tacks in the end bars. You will soon learn to judge how much current and pressure are needed on the wires before they sink into the wax. To begin with you may end up with a sheet of foundation sliced neatly into several pieces. If this happens take one probe off but continue to hold the frame down — if you hold the edges of the foundation together they may rejoin. With practice you can embed foundation quickly and with few errors.

Lids

Telescopic lids

Assembling bought lids is easy using 45 or 50 mm galvanised flat-head nails. Home-made lids should provide sufficient clearance over a standard hive box (505 x 405 mm) to allow for easy removal, with internal measurements of the lid about 510 x 410 mm.

A sheet of galvanised steel (up to 735 x 635 mm) is usually bent over the lid and fastened with screws or galvanised flat-head clouts through pre-drilled holes. Recycled aluminium alloy sheets from offset printing works make much cheaper roof cladding. They are easier to bend than galvanised steel, but less robust.

Plastic lids and galvanised lids with sprung ends do not require assembly.

Migratory lids

These do not have overlapping sides, though some migratory lids do have end battens 50–75 mm deep. Migratory lids rely on the bees propolising them firmly in place, so inner covers or hive mats should not be used under them. Some beekeepers fit small triangular pieces of flat galvanised steel to each corner by nailing them to the top decking and the end battens. This strengthens the end battens and also prevents the lid from being skewed off the hive.

The lid decking can be made from timber, but a much cheaper alternative is 12 mm construction-grade plywood. Plywood also does not warp as much as 20 mm timber. Ten lids can be cut from a 2400 x 1200 mm sheet. Plywood lids should be treated with a fungicide, and then painted thoroughly on both sides. Galvanised steel can be folded to cover the lid decking.

Bottom boards

These can be made from scrap timber, and dimensions are not critical. Use galvanised nails for assembly. Use large nails or screws to fix the floorboards to the runners, so they don't fall apart when hives are being shifted and dragged along a truck or trailer deck.

Other equipment

Inner covers and queen excluders are assembled simply by nailing the four pieces of the rim together. To add extra strength to the corners glue the joints and/or nail triangular-shaped pieces of flat aluminium or galvanised steel across each corner.

PRESERVING HIVE EQUIPMENT

Taking care to preserve hive parts will ensure that they have a long life — well-preserved honey boxes should last 20–30 years or more, and brood boxes 10–15 years. The best preservation method that does not require expensive equipment is to use a fungicide solution, and then paint the hive.

Fungicides

Hive parts exposed to weather should be treated with a fungicide such as Metalex®. Most concentrated fungicides can be diluted with kerosene, turpentine or paint thinners according to the manufacturer's instructions. Copper sulphate (bluestone) from garden shops can also be used as a 5 per cent solution dissolved in water. Fungicides can be applied by brushing or rolling, but it is best to soak hive parts for some hours before they are assembled.

After you have assembled treated equipment, leave it for several weeks before painting, especially if using kerosene as the diluent, as it is slightly oily. Equipment can also be purchased pre-treated with copper-based products such as Tanalised® Ecowood™ (Tan®E).

A new timber product now available on the New Zealand market is called Thermo-Wood®. This timber does not contain any chemicals, but has been heated to extreme temperatures, which modifies the hemicelluloses or wood sugars to make them unavailable to timber-destroying fungi. The timber is very light but somewhat brittle, and holes may need to be pre-drilled before nailing.

Paint

There is no substitute for good surface preparation and good-quality paint. Water-based and oil-based paints are equally effective, but oil-based paint doesn't breathe and water passing through the timber from inside the hive box can blister the paint. Water-based paints are also much easier to use.

Use an exterior-grade primer then two coats of water-based house paint, or self-priming paint, according to manufacturer's directions. Paint only the outside and rims of boxes, as it is neither necessary nor desirable to paint the inside.

The colour of boxes is unimportant, and you can save money by buying specials and returned colours. Doing this is actually an advantage, because a variety of colours on different boxes helps the bees to find their own hives. Lighter shades are preferable because they attract less heat from the sun, so last longer than darker colours. Bees can differentiate between yellow, blue, and blue-green, although other colours may appear as different tones to them. Bees cannot see the colour red, which appears to them as a neutral 'grey'.

Paraffin dipping

One method of preserving hive parts that has been used in New Zealand for a long time, and seems to have been developed here, is dipping in hot paraffin wax. Most commercial beekeepers own a paraffin dipper, and hobbyists can either use one on a contract basis or pool their resources to have one built.

There are many different types of wax dipper, from the simple one described here (Fig. 3.15) to large push-through models capable of treating boxes at a fast rate. The unit described here has the following features:

- a steel vat to contain the paraffin wax
- a brick surround to provide a double skin for the vat
- a 2 m chimney for good draught
- a firebox door to enclose the fire, for safety reasons.

Some beekeepers heat the vat using a gas ring and LPG. Gas heating has the advantage of being cleaner, and less of a fire danger if the wax overflows.

Fig. 3.15 Simple paraffin dipper with a brick surround and tall chimney.

Fig. 3.16 Paraffin dipper with a central chimney, and a handle for holding boxes below the surface of the hot wax.

The vat sits in the brick surround, supported by a 50 mm lip of 5 mm plate steel on all sides. The vat in this case is 635 mm long, 430 mm wide, and 480 mm deep to the rim, sufficient to hold two three-quarter-depth boxes and still leave some freeboard. The actual dimensions will depend on the type of hive equipment used.

The bottom is made of 6 mm plate steel, and the sides are 3 mm thick. All joints are welded from both sides to reduce the risk of bursting when in use. The vat hangs inside the brick surround with about a 30 mm gap on all sides to allow the fire to circulate. This gap is not so necessary if gas is used.

At the rear of the dipper there is an extension that accommodates the chimney. Bricks are used as a surround because they radiate much less heat than a steel-walled firebox, making the job of wax dipping a lot more comfortable. A grate is provided in the firebox, and while this is not essential it does make managing the fire easier. For safety reasons the firebox is enclosed with a door, such as one from an old copper, although a small hole must be chipped in the mortar below the door to allow air into the fire. The 2 m chimney draws well, and keeps smoke away from those using the dipper.

Similar dippers are often made completely of plate steel (Fig. 3.16), rather than having an outer skin of bricks.

Paraffin wax is available from oil companies and beekeeping equipment stockists — the grade with melting point around 60 °C is best. This melter holds 85–90 kg when filled to its normal operating level 125 mm below the rim. The wax should be heated to 150–160 °C, a process that may take a couple of hours. It is a good idea to use a candy thermometer attached to a stick and placed in the wax, rather than relying on guesswork. At the correct operating temperature, the wax really bubbles and boils up around the boxes when they are immersed. If the wax is too hot or the timber is too moist, the wax can froth up over the edge, or the surface can catch on fire, either way with potentially disastrous results.

Before being dipped, all woodware must be as dry as possible. The wax tends to seal any water in the timber, and

Fig 3.17 Painting a box fresh from the paraffin dipper, using a rotating cradle to make the job easier.

excess water leads to violent frothing and spilling of the wax. Immerse the hive parts for several minutes, using a heavy weight such as another box or brick to keep them below the surface. Then lift them from the wax onto a draining board — this can be permanently fixed to the dipper or can be simply made out of a sheet of corrugated steel. Put two more boxes in the wax. By this time the draining supers will be free of surplus surface wax, and are ready for a quick scrape to remove propolis and any flaking paint. Boxes previously painted with oil-based paints usually blister and will need a good scrape.

Boxes can be used straight after dipping without being painted, but while paraffin wax is good protection on its own the paint protects the boxes from sun damage as well as giving them a neat and tidy appearance. You should paint paraffin-dipped boxes while still hot, with two quick coats of water-based paint (Fig. 3.17). Put on thick coats, as the paint will dry very quickly and will be pulled into the wood by the cooling wax to give a very good 'fix'. If the boxes are allowed to cool too much, water-based paints will not adhere.

Woodware is very hot when it comes out of the dipper, so use tongs or thick rubber gloves. It helps to put a little cold water in the gloves first.

Paraffin dipping gives boxes a very effective protection against rot that will last for a long time, even in high-rainfall areas. The cost per box is not great — 1 kg of wax will treat four or five new full-depth supers or six previously painted ones.

Finally, a few safety precautions:
- locate the wax unit away from buildings
- wear some eye protection in case of wax splashing
- have a metal cover or damp sacks handy to put over the vat in case the wax boils over or catches fire
- regularly scoop rubbish from the bottom of the dipper if refurbishing old boxes
- regularly skim the surface of the wax to remove debris, and scrape the edge of the dipper at the wax level to prevent fires
- never put damp woodware or water into a dipper
- empty any water out of the vat before starting the fire.

TOXIC PRODUCTS

All grades of tanalised timber contain arsenic, which is highly toxic to bees. This timber should not be used for hive parts, except where indicated earlier in this chapter. However, the product called Tanalised® Ecowood™ (Tan®E) contains only copper azole, and not the chrome or arsenic compounds typical of other tanalised products.

Interior-grade boron-treated timber (usually pink in colour) is not known to be toxic to honey bees, but neither does boron treatment give any lasting protection against rot. Treated timber with hazard-class descriptions H3 to H6 contains preservatives that are toxic to bees, and should not be used for hive parts.

Some plywood may contain insecticide-impregnated glue, and is marked TGL (treated glue line). This is safe to use for lids and floors if it is well painted on both sides.

Particle board and fibre board are not suitable for hive parts, as they swell and disintegrate when wet. Many may contain urea-formaldehyde glues, which release fumes toxic to bees.

As you become established as a beekeeper you will invest a lot of money in hive equipment. Using good-quality equipment, and paying careful attention to assembly and preservation, really pay off in the long run.

Honey bee biology  4

To keep bees successfully you need to learn at least the basic details of honey bee biology — how bees develop, how they function as individuals, and how the colony behaves as a unit. Honey bees are among the most fascinating animals on earth, and many people start beekeeping because they are more interested in the bee than they are in its products.

This chapter looks at honey bee biology from a beekeeper's point of view, so that you can understand why beekeeping is practised in particular ways rather than just learning techniques by rote. There's a wealth of further information in some of the specialised books listed at the end of this book.

In New Zealand beekeeping means the culture and husbandry of the honey bee *Apis mellifera*. We've included information about some of the honey bee's relatives in the latter part of this chapter. Bumble bees, alkali bees and leafcutter bees are also farmed and used for pollination in a small way in New Zealand.

HONEY BEES

Of the seven or more species of honey bees in the world, one has been introduced to New Zealand. This is the western honey bee, *Apis mellifera*, the most important bee species to be managed on a commercial basis for pollination and high honey production. In some countries the smaller Asian honey bee (*Apis cerana*) is farmed in hives like the western honey bee.

Many subspecies of the honey bee have been described, and a number have been imported to New Zealand. The bee stock here is predominantly Italian, with Carniolans gaining in popularity and some European black bees still surviving the ravages of varroa. Importation of honey bees is strictly controlled. The only material that has been allowed in during the last 50 years is the semen of Italian honey bees imported from Australia in 1988 and 1989, and Carniolan semen from Austria and Germany beginning in 2004.

Fig. 4.1 A marked queen honey bee surrounded by worker bees.

Fig. 4.2 Top: Drone and worker.
Bottom: Queen and worker.

Members of the honey bee colony

The population of a honey bee colony is made up of three distinct types, called castes — drones, workers and queens. The males (drones) are quite different in body form from the females (workers and queens). The first step in learning about honey bees is being able to distinguish between the three castes, which are easy to tell apart after some practice.

The abdomens of queens and workers are both pointed, but differ mainly in their size (Fig. 4.1). A worker's abdomen is short, and is nearly covered by the wings when at rest. A mated queen bee has a long abdomen, little more than two-thirds of which is covered by the wings when at rest. Her abdomen is much longer because it contains developed ovaries, where the eggs she lays are produced.

Drones have heavy, stout bodies with blunt abdomens (Fig. 4.2). Their large compound eyes meet at the top of the head, rather like those of a blow fly. Queens and workers (Fig. 4.2) both have areas of hair between their eyes.

The three castes go through the same stages of egg, larva, pupa and adult, but take different times to develop (Figs 4.3, 4.4).

Fig. 4.3 Development of worker bee. Top L to R: egg, four larvae of increasing age. Bottom L to R: four pupae of increasing age.

Day	Worker Stage	Worker Moult	Queen Stage	Queen Moult	Drone Stage	Drone Moult
1	EGG		EGG		EGG	
2						
3		(hatching)		(hatching)		(hatching)
4	LARVA 1	1st moult	LARVA 1	1st moult	LARVA 1	1st moult
5	LARVA 2	2nd moult	LARVA 2	2nd moult	LARVA 2	2nd moult
6	LARVA 3	3rd moult	LARVA 3	3rd moult	LARVA 3	3rd moult
7	LARVA 4	4th moult	LARVA 4	4th moult	LARVA 4	4th moult
8		(cell sealed)	LARVA 5 or PREPUPA	(cell sealed)		
9	LARVA 5 or PREPUPA					
10				5th moult		(cell sealed)
11		5th moult				
12			PUPA		LARVA 5 or PREPUPA	
13						
14						
15						5th moult
16	PUPA		ADULT	6th moult (emerging)		
17						
18					PUPA	
19						
20		6th moult				
21	ADULT	(emerging)				
22						6th moult
23					ADULT	(emerging)
24						

Fig. 4.4 Developmental stages of the three honey bee castes.

Queens

Honey bee colonies will rear queens under three different sets of conditions (called queen-raising impulses):

- Supersedure takes place when an old queen is eliminated and replaced by one raised within the same colony. The old queen is still within the hive when the new one emerges, and may even continue laying eggs for some time before eventually disappearing.
- Swarming is the means by which the colony reproduces and disperses. When a colony swarms, new queens are raised, but before the first one emerges the old

queen flies from the hive with a proportion of the workers to establish a new colony. Some of the remaining virgin queens may also leave with a small cluster of worker bees, and these are known as afterswarms. Eventually one of the queens reared in the parent colony flies and mates, returns to the colony, and any remaining queens or queen cells are destroyed.

- The third queen-raising impulse is the emergency one. If a colony's queen suddenly dies a new queen can be raised, by workers modifying some worker cells and the diet of the larvae in them.

Queens begin their development as fertilised eggs, laid in queen cell cups prepared by worker bees (swarming and supersedure), or in worker cells (emergency impulse). The eggs are 1.5 mm long, white and sausage-shaped. After three days a larva hatches from the egg and is fed copious quantities of royal jelly, a rich glandular secretion of worker bees.

Fig. 4.5 Newly emerged virgin queen, and queen cell with neatly opened end.

The queen larva is fed royal jelly for five days, until the cell is sealed with wax and the larva spins a cocoon before pupation. The pupal stage lasts six days, after which the fully developed queen emerges by chewing the cap off her queen cell. The beekeeper can usually see whether a queen has emerged from a cell, in which case a neat cap will have been removed from the end (Fig. 4.5). If the cell has been torn down from the side, the queen has been killed by other bees.

When a virgin queen emerges from her cell she has a fairly short abdomen, a feature that distinguishes her from a mated queen. She begins to fly from the hive within a few days, making one or two short orientation flights that enable her to memorise the location of the hive and nearby landmarks. Mating with drones takes place in the air away from the hive, and the virgin goes on one or several mating flights — within about six to ten days of emergence in summer or autumn, but anything from two to four weeks after emergence in spring. These flights usually take place on a warm afternoon, and may range up to several kilometres from the hive.

Queens mate with an average of 12 different drones (usually from other hives) but the mating number is highly variable. Some five or six million sperm from these matings migrate to an organ called the spermatheca. In a well-mated queen the supply of sperm stored there is sufficient to fertilise her eggs for the several years of her life.

A newly mated queen returning to the hive may still carry the reproductive parts of the last drone she mated with. This 'mating sign' is soon removed by hive bees. After mating, the queen's ovaries expand and her abdomen fills out, so she takes on the characteristic appearance of a mated queen. The queen will not fly out of the hive again, except perhaps with a swarm.

Egg-laying may start as soon as 12 hours after a queen returns, but most queens begin laying in two or three days. A vigorous queen at the height of the season will lay an average of approximately 1500 eggs per day, equivalent to her own body weight. This high output is maintained by the workers constantly feeding her royal jelly. The bees around the queen also groom and lick her, removing excrement and any eggs that are accidentally dropped.

A queen's egg-laying rate is determined by the workers. They control the amount of food given to her and the number of cells cleaned out and prepared for an egg. Before laying in a cell, the queen inspects it and measures its size. In larger cells (drone cells) she lays unfertilised eggs, which develop into drones. In smaller worker cells and in even larger queen cells, she lays fertilised eggs that normally develop into females (workers or queens). This determination of the sex of the offspring continues with almost unfailing regularity until the queen begins to lose productivity in old age.

Queen bees also produce several chemicals collectively called the 'queen substance'. These act as pheromones — chemicals produced by one individual (in this case the queen) that affect the behaviour or physiology of other individuals (workers, drones and other queens). Queen substance is composed principally of two products of the mandibular glands, called oxodecenoic acid and hydroxydecenoic acid. The substance is removed from the queen by the workers that groom her, and passed around the colony as workers exchange food.

Queen substance inhibits the workers from constructing queen cells, prevents the workers' ovaries from developing, and otherwise holds the colony and swarms together as a coherent unit. It is also a sex attractant and worker attractant.

Queens in most colonies generally live for one to three years, with few queens lasting longer than that. There are, however, records of queens living for much longer, especially if kept in nucleus colonies rather than production units, and if the beekeeper continually removes swarm cells.

Drones

Drones are male bees whose main function is to mate with queens. They apparently have no duties in the hive and do not forage, and they lack pollen baskets, sting, and wax glands.

Drones normally develop from unfertilised eggs laid by the queen in larger (6.5 mm diameter) cells called drone cells. This development from unfertilised eggs is called parthenogenesis or asexual reproduction, a process that is also common among other types of insect. Because they originate from unfertilised eggs, drones have only half the number of chromosomes per body cell that workers or queens do (16 as opposed to 32).

Drones take 24 days to develop to adult stage: three days as an egg, six days being fed as a young larva, and 15 days in a sealed cell as a larva and a pupa.

A virgin queen, or an old queen that has run out of sperm (drone layer), both lay drone eggs,

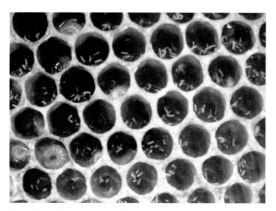

Fig. 4.6 Multiple eggs in worker cells, from laying workers.

but will do so in worker cells as well as drone cells. If a colony becomes queenless, some workers may develop functional ovaries and begin to lay eggs. These laying workers cannot mate and so will only produce drone eggs. Their eggs will be laid mainly in worker cells (Fig. 4.6), but the brood will have the characteristic domed cappings of drone brood. Drones from eggs laid by workers will usually develop to maturity, but are smaller than normal.

Young drones are fed a mixture of brood food, honey and pollen by workers for the first few days of their adult life, and later feed themselves with stored honey. Drones begin to fly at about eight days old, at first taking short flights to orientate themselves to the hive. They are sexually mature and capable of mating after about 12 days, and on most warm afternoons they take flights of 30–60 minutes in search of flying queens. Drones may congregate in areas, or fly for several kilometres from the hive at a height of 10–40 m above the ground, finding virgin queens by following their pheromone trails and then by sight.

The large eyes of a drone are particularly sensitive to movement, and a patrolling drone that has been aroused by queen pheromones may even chase after other flying insects or a pebble thrown into the air. When mating, a drone mounts a queen and then everts its penis and mates. The penis is torn from its body when the drone falls from the queen after mating. Drones die within minutes or hours of this taking place.

Drones that do not mate usually live for about three to four weeks in spring or mid-summer, though up to about three months in late summer and autumn. With the shortening of daylight and reduction of food supplies in autumn, they are normally expelled from the hive and die. Drones are reared again in spring.

A plentiful supply of drones in spring and summer is very important to the survival of the honey bee species, and beekeepers are wrong to think of them as being unnecessary in the colony. Squashing drone brood and culling every last scrap of drone comb from a hive are particularly futile actions, as bees will expend a lot of energy to repair the damage. Colonies in hives with plentiful drone comb do not produce less surplus honey than those with no drone comb, and bee colonies seem to require a certain percentage of drone brood.

As drone brood is often preferentially parasitised by varroa, some beekeepers cull drone comb. This may be a mistake, as one of the reasons for observed lack of longevity of queen bees in the presence of varroa could be poorer matings because there are fewer drones. The widespread adoption of plastic foundation is also reducing the amount of drone comb in hives, as plastic combs don't suffer the same damage as wax foundation combs. Colonies usually repair damaged comb by building drone cells.

Workers

The vast majority of colony members are worker bees, up to about 50,000 in mid-summer but much smaller numbers during winter. Workers are females, having ovaries, but are non-reproductive as they cannot mate and do not normally lay eggs.

A worker bee develops from a fertilised egg laid by the queen in the base of a small (5 mm wide) worker cell. After three days a larva hatches and is immediately fed brood food from the glands of adult workers. After two or three days of being fed brood food, lesser quantities of pollen and honey or nectar are added to the diet. During the total five days of feeding, a worker larva grows to 240 times its original weight, moulting regularly and passing through five larval stages or instars.

Fig. 4.7 Worker bees feeding royal jelly to queen cells.

On the ninth day after the egg is laid, adult worker bees seal the larva in its cell with a wax capping. No feeding can take place after this. The larva (sometimes called a prepupa at this stage) spins itself a rudimentary cocoon, inside which it spends nine days pupating.

During the pupal stage, distinct adult parts such as wings, legs, mouthparts, antennae and eyes begin to appear. The pupa gradually changes from white to its adult colours, and hair also develops. On the twenty-first day the young adult worker bee chews a hole in the cell capping and emerges (Fig. 4.8). It takes a feed of honey, or is fed by other worker bees, and grooms itself.

The life span of adult workers is extremely variable and depends on the activities carried out by the bee, the conditions at the time and the food available. In temperate

Fig. 4.8 Young worker bee emerging from cell.

climates workers are generally reckoned to live for a short time in summer (averages of 15–38 days have been recorded), a somewhat longer time in spring and autumn (30–60 days), and much longer in winter, perhaps an average of 140 days. Long-lived 'winter bees' have better-developed hypopharyngeal glands and fat bodies because of higher pollen consumption in the autumn. They are also less active and carry out much less brood rearing and wax secretion.

Activities and behaviour of the honey bee colony

Workers carry out almost all the tasks inside the hive, and generally do so in a sequence called 'division of labour' that is related to their age. Many studies, some going back hundreds of years, have built up a picture of this age-related pattern of work. Much of the pattern is due to the different times at which the glands used for feeding brood, secreting wax and producing enzymes to process honey mature.

As attractive as a rigid schedule of activities is, the evidence shows that the picture is much more complicated. There is a huge variation in the ages of bees carrying out any particular task, and individual bees can perform a number of different activities at any given age. As the needs of the colony change, worker bees can revert to doing tasks they had done at a younger age.

Also complicating any analysis of worker activity is the fact that workers spend most of their time doing nothing, or simply patrolling about the hive. The much-lauded 'busy bee' appears in fact to be rather lazy, though patrolling probably is an important part of assessing the colony's needs and providing a mobile workforce for tasks that need doing.

The division of labour so often talked about does exist, but is a flexible system in which workers go through a series of activities starting with nest tasks and ending outside the hive. The pattern of this sequence is modified by the colony's needs and seasonal conditions.

Nest cleaning

The first activity of a worker after it has emerged is often cleaning cells, as remains of larval cocoons and excrement must be removed before the queen can lay in a cell again. Other nest-cleaning duties include removing cappings from cells and taking away debris from the hive such as cappings and dead brood or adults.

Caring for brood

An important activity of worker bees inside the colony is caring for brood ('nursing'), especially feeding it. Visits to brood cells by nurse bees start as soon as an egg is laid, and feeding begins once the egg hatches. It was widely thought that older bees (with better-developed glands) fed younger larvae, and that younger bees fed pollen and honey to older larvae, but closer studies have shown no age difference between bees visiting larvae of all ages.

Not all visits to brood result in feeding — presumably the high number of visits helps with temperature regulation inside the cells. The number of visits each larva receives varies with colony conditions, but is probably at least 2000 over the five days it is uncapped, including 150 feeding visits. The composition of the food given changes as the larva develops, indicating that nurse bees can detect the age of larvae. This change in food composition is essential in ensuring the larva develops into a worker, and not into a queen.

Tending the queen

The queen is normally surrounded by a group or 'court' of attendant workers, up to ten or so bees examining the queen, licking her body and feeding her brood food and honey. The grooming process removes pheromones from her, and as the workers in the court are replaced frequently, these chemicals are quickly transmitted through the colony by worker-to-worker feeding.

Comb building

A worker bee has four pairs of wax glands on the underside of its abdomen, which are usually most active from eight to 17 days old. Comb building is a communal process, not an individual one, and before it begins there must be a high temperature (33–35 °C) created by the bees within the hive, enough worker bees with active wax glands, and an abundant supply of food. Wax secretion is energy-demanding, as between 8 and 13 kg of honey are required to produce 1 kg of wax.

Worker bees about to secrete wax gorge themselves with honey, and hang together in clusters or chains across the space to be filled with comb (Fig. 4.9). After about 24 hours, wax is produced by the glands. A liquid at first, the wax soon hardens to a small white flake. Using spines on the hind legs, the bee removes each flake in turn and passes it to the mandibles (jaws), where the flake is chewed and mixed with glandular secretions. The manipulated wax flake is normally passed to other workers, which mould it into shape and push it into place on the growing comb. A single flake takes about four minutes to process.

Fig. 4.9 Worker bees joined together in chains, building new comb in the hive.

Fig. 4.10 Natural worker comb.

Bees build combs vertically by responding to hair-like gravity sense organs in their antennae and joints. The direction in which natural combs are built is determined mainly by the shape of the cavity occupied by the bees, but it is interesting to note that a swarm arriving in a perfectly regular cavity may use the earth's magnetic field to orientate their combs to the same direction as those in the nest just vacated.

The bees' comb (Fig. 4.10) is used for both brood rearing and food storage. The tightly interlocking hexagonal shape of the cells gives strength and economy of space. Combs hang vertically in the hive with individual cells nearly horizontal, sloping slightly upwards to prevent larvae and honey falling out.

Beeswax is an amazing substance. The wax used to make the base of the cells of natural worker comb is 0.18 mm thick and the walls are only 0.07 mm thick. A piece of worker comb weighing 500 g when empty will be made up of about 35,000 cells, and can contain 10 kg of honey. The cells at the top of a piece of natural comb 300 mm deep and attached only at the top, which is full of honey, will be supporting over 1300 times their own weight. A square of comb 100 x 100 mm contains 429 worker cells or 260 drone cells on each side. Natural worker brood combs are about 34 mm centre to centre, while drone combs are 38 mm between centres.

Guarding

A hive's entrance is open to all animals small enough to fit through it but is guarded by certain worker bees, many of which are about two-and-a-half to three weeks old. Guard bees wait on the hive's landing board or just inside the entrance, usually standing on

Fig. 4.11 Guard bees attacking a wasp at the hive entrance.

the mid and hind pair of legs with their forelegs raised and antennae forward (Fig. 4.11). Guards are alert and often inspect incoming bees with their antennae to detect their colony odour — a mixture of chemicals unique to each colony.

During a honey flow, guard bees don't usually inspect incoming foraging bees from their own hive or bees that have drifted accidentally from another hive. When there isn't a flow the guarding is more rigorous. Bees, wasps and other intruders seeking to 'rob' honey from the hive travel with a different motion as they approach, which alerts the guard bees to inspect them. Intruders are usually stung by the guards, and the alarm pheromones (including isopentyl acetate and 2-heptanone) recruit other workers to assist. The initial number of guard bees may be small, but the pheromone communication system ensures that serious invasion is dealt with rapidly.

Scenting

When a hive is disturbed — for instance when a beekeeper pulls it apart — many bees fly that have never before flown from the hive and so don't know its location. At such times workers can be seen standing in front of the hive with their abdomens raised and their wings beating rapidly.

These bees are producing orientation pheromones that attract the flying bees back in. Worker bees releasing these pheromones (or 'scenting') have their abdomens raised with the tips curved down. You may be able to see a small brown gland exposed between the skeletal plates near the tip of the abdomen (Fig. 4.12). This is the Nasonov gland, source of the orientation pheromones. You can also see scenting behaviour when a swarm is hived, and bees stream into their new abode.

Fanning

Through metabolic activity, bees produce a lot of heat in the hive, and during warmer periods hot air is removed by bees at the hive entrance fanning their wings. Fanning is particularly obvious when a lot of nectar is being ripened into honey, as much water is removed from the nectar and the bees reduce the relative humidity inside the hive. Much of this activity takes place at night, with bees fanning in the

Fig. 4.12 Bees scenting with their Nasonov glands exposed.

cooler air with lower humidity and expelling more humid warmer air. On hot days bees also evaporate water from brood cells to cool the hive.

Fanning bees may be seen at the hive entrance, gripping the base board with their abdomens curved downwards and fanning their wings vigorously. Usually they are tail-outwards to expel air, but on especially hot days or when honey is being ripened, some bees can be seen tail-outwards on one side of the entrance and tail-inwards on the other, to circulate air within the hive.

When the hive interior is very hot or humid many of the bees cluster at the front, or 'hang out' as beekeepers say. This problem can be alleviated if extra ventilation is provided by chocking up the front of the hive. Bees hanging out normally retreat into the hive overnight.

Orientation flights

Worker bees may first fly when about a week old, but take only short flights near the hive. These orientation or 'play' flights allow the bee to memorise the location of the hive and nearby landmarks. After several days of bad weather many young workers take their orientation flights together, and you may notice this during the middle of the day as a cloud of bees milling slowly around the hive. This isn't the same as swarming, though you may confuse the two at first. Orientation flights last for only a few minutes, while bees about to swarm leave the hive in a rush and head off rather than mill around the hive.

During orientation flights young bees may drift to another hive if the apiary is laid out with few visual clues for them to use.

Communication associated with food sources

Much of the early research on honey bee communication was carried out by the German naturalist Karl von Frisch, who published his first work on the subject in the 1920s. Von Frisch studied many aspects of honey bee perception and communication, and most of his discoveries are summarised in *The dance language and orientation of bees*. He shared a Nobel Prize for his work in 1973, and died in 1982.

Von Frisch determined how bees communicate through movement, or 'dance'. He described two main types of communication dance, which he called the round dance and the wag-tail dance.

The round dance is performed by scout bees that have found a food source less than about 100 m from the hive. The scout bee runs in small circles on the comb, rushing clockwise and anti-clockwise with quick, short steps. It dances at one place on the comb for anything from a few seconds to minutes, and then moves off to dance at another spot. Bees near the dancing bee become excited and may follow it in the dance before leaving the hive to search for the food source.

The dance gives no information on the direction of the food, and bees search anywhere within about a 100-metre radius. This general searching is why robbing in urban apiaries is so serious. Foragers are told of a very good food source somewhere within 100 m of the hive and fly out in all directions looking for it, causing a nuisance to neighbours in the vicinity of the apiary. This can also happen when extracted supers are put on a hive, or hives are fed, during the day.

If the source of food found by a forager is more than about 100 m from the hive, the returning bee performs the wag-tail dance. This dance communicates the distance and direction of the food source to the bees that gather around the dancing bee.

The dance consists of a straight run in which the bee's abdomen is vibrated or waggled, followed by a half-circle to begin the straight run again. The returning path from this next straight run is made on the opposite side, so giving the dance its figure-of-eight pattern.

The distance to the food source is communicated by the length of time spent in the straight run — the further away the food source, the longer the straight run. The direction taken to reach the food is given relative to the sun's position in the sky, but inside the hive gravity is used as a reference point for this information. If the food source is directly towards the sun, the straight run is performed directly upwards on the comb. A source directly away from the sun is indicated by performing the dance straight down on the comb, and all other directions can be given relative to these reference points.

Other information communicated in the dance includes the scent and sugar concentration of the nectar at the source, as the dancing bee stops periodically to share out nectar with attendant bees. The quantity of food available is indicated to the colony by the number of dances being performed in a day. Bees that quickly gather a full load of nectar or pollen are able to perform more dances in the hive, and thus recruit more bees to that source.

Many plant species yield nectar or pollen only at particular times of the day. These times are indicated by when the dance is performed, and are remembered quite accurately by the bees for many days. This saves them making wasteful visits to the source when nectar and pollen are not available.

Several potential problems associated with using the sun as a reference point are neatly overcome by honey bees. The sun's movement through the sky and its seasonal variation would result in new foragers missing the food source by a wider and wider margin, were it not for a 'sun-compass reaction' that automatically compensates for these changes. The sun is not always visible to humans, but on a cloudy day honey bees use the plane of polarisation of sunlight in any patches of clear sky to calculate the position of the sun. Their eyes are also sensitive to ultraviolet light (which human eyes are not), and as some UV passes through clouds bees use it to determine where the sun is.

The dance communication system has built-in safeguards that, for instance, prevent bees being recruited to forage on toxic nectar. A fast-acting poison would kill a forager before it could return to the hive, a slow-acting one might cause it to get lost on the way back, or a poison might be strong-smelling enough for the guard bees to reject the forager at the hive entrance. Normally foragers make several trips to a source before they dance, and hive bees need to have the dance performed several times before they are recruited, which gives extra safeguards.

Further study of the honey bee's communication system will uncover more about the types of information communicated in the dance. It is currently thought that details of the colour and shape of flowers are not communicated, and it is generally held that information on the elevation of the food source is not given, although there is some doubt about this.

As neat as the von Frisch dance communication model is, it is not universally accepted. Other scientists, notably Wenner and Wells, argued strongly that the bees used their sense of smell almost entirely to locate food sources. It may well be that dance communication is used to recruit bees to a general area, and that odour plays a more important role in the bees homing in on the exact source of nectar or pollen than von Frisch's model suggests.

Whatever mechanism or mechanisms bees use, they are amazingly successful at finding a food source and recruiting other foragers. Bees also dance for pollen and water sources, as well as nectar, and scout bees in swarms dance to indicate the position of potential nesting sites.

Collecting and storing food

At about three weeks of age a worker honey bee begins to forage for nectar, pollen, water or propolis — depending on the needs of the colony at the time. Only a very small number of bees forage for water or propolis, and generally bees will collect either nectar or pollen, or both. Workers often specialise in collecting either pollen or nectar for quite some time, but can change tasks as the colony's needs alter.

When on a foraging trip from the hive, almost all worker bees will collect either nectar or pollen from one plant species only. This very high 'crop constancy' is one of the reasons honey bees are such good pollinators, as pollen from one species is not wasted by being transferred to the flowers of another. Other pollinating insects, such as bumble bees, do not behave this way.

A worker bee will continue to forage on a particular species as long as it is a good food source, returning to the hive with a full honey sac and/or pollen baskets, and dancing to recruit new foragers to the source. It is quite possible that a worker bee will spend all its foraging life on one plant species.

Some bees are not recruited by dancers to forage on a particular crop, but instead go from the hive as scout bees to search for new nectar or pollen sources. On finding a good food source they communicate its location to foragers, and later may even become foragers fixed on that particular food source. Not all the hives in an apiary necessarily forage on the same crop, especially for pollen. You can see this very clearly if you have pollen traps on several colonies in an apiary.

Most foragers fly from the hive when the air temperature is higher than 12–14 °C and the wind less than 25 km/h. During the flight from the hive to the food source their speed ranges from 10 to 30 km/h, with an average, according to some studies, of 23 km/h. The speed of returning bees is faster and less variable, ranging from 21–26 km/h with an average of 24 km/h.

Pollen collection and storage

Pollen is the reproductive stage of flowering plants and is the source of protein, minerals, fats, vitamins, and trace elements that are necessary for the growth of larvae. It is vital for the production of brood food by worker bees, which is used to feed larvae. A colony will collect anywhere from 15 to more than 50 kg of pollen per year.

A pollen-collecting bee behaves quite differently on a flower from a nectar collector, walking and scrabbling all over the flower to collect pollen on its hairy body.

The pollen is brushed off with the legs in a complicated set of movements. The first pair of legs gathers the pollen deposited on the head, mouth-parts and antennae. The middle pair of legs collects pollen from the thorax and takes that gathered by the first pair of legs. The pollen now on the second pair of legs is removed by the 'pollen comb', a special structure on one of the joints of each hind leg. The pollen is then transferred to the pollen basket or corbicula, a flattened area on the tibia of each hind leg with a single hair around which the pollen is pressed (Fig. 4.13). This hair keeps the pollen in place during the flight back to the hive.

The amount of pollen collected on each flight, and the time taken for a bee to get a full pollen load, depends on the flowers being visited. The two pollen pellets together weigh between 10 and 30 mg, and result from visits to anywhere between one and several hundred flowers. Pollen collectors probably average about 10 round trips per day.

Fig. 4.13 Foraging bee with pollen dusted on the body and packed into the pollen baskets.

When a full load has been gathered the pollen-collecting bee returns to the hive. It may dance on the face of the comb, and it then finds a suitable cell in which to store the pollen. The two pellets are simply dropped in the cell, and are later mixed with nectar and compacted by hive bees. The pollen may be covered by a layer of honey. The pollen and nectar mix undergoes lactic acid fermentation and becomes what is called 'bee bread'. In this form it is preserved and can remain nutritious for many months.

Nectar collection

Nectar is a sweet liquid secreted by plant nectaries, which are usually located in flowers. It is composed almost entirely of sugars and water, and provides bees with a carbohydrate energy source.

Bees are attracted to flowers by their colour, odour, shape and movement. A nectar-collecting bee will insert its proboscis (mouth parts) into the nectary to check for nectar (Fig. 4.14). Any nectar is gathered, and several hundred flowers are usually visited before a full load (averaging 25–40 mg) is stored in the bee's first stomach — the proventriculus or 'honey sac'.

If the nectar flow is weak, the forager will return to the hive, regurgitate its load of nectar, give it to several hive bees and

Fig. 4.14 Foraging bee collecting nectar.

then return to the field. But if the flow is strong, the forager will return to the hive and dance on the face of a comb among other workers. It will stop periodically to offer some of the nectar to other potential foragers, and after a period of dancing it will give its load of nectar to hive bees, clean itself, have a feed of honey and return to the field.

Robbing

Robbing is the name given to a particular type of foraging behaviour where bees take honey from another hive or from open honey boxes, or collect exposed sugar syrup. If given a choice, bees will forage for nectar rather than honey, but will take honey when nectar becomes unavailable.

Scout bees that find a honey source quickly communicate its location to foraging bees. The number of bees robbing a hive increases rapidly and a robbing frenzy may soon develop (Fig. 4.15). The air becomes filled with bees flying backwards and forwards around the hive, seeking an entrance to get at the honey. The nervous flight of robber bees is unlike that of returning foragers, which make straight for the hive entrance, or the orientation flight of young bees.

Severe robbing can destroy a bee colony as well as cause bad relations with neighbours. You should be very careful to prevent it from happening, and you should know how to control it if it starts — these issues are discussed in Chapter 14.

Fig. 4.15 A hive under severe robbing attack, with invading bees flooding the entrance and trying to get in at other places.

Ripening nectar and storing honey

Hive bees are responsible for processing the nectar collected by foraging bees into honey and storing it in the hive. Two processes take place when nectar is transformed into honey — water is removed and the sugars are changed chemically. These two processes are collectively called 'ripening' the nectar.

Nectar from different plants varies in sugar concentration, but the sugar is usually the double sugar molecule (disaccharide) sucrose, the same as white table sugar. Bees convert the sucrose to the two simple sugars (monosaccharides) glucose and fructose, by adding the enzyme invertase. This enzyme is produced in the hypopharyngeal glands of older hive bees and foraging bees, the same glands that produce royal jelly in younger bees. The invertase is added to nectar, first by the returning forager and then by hive bees, to speed up the conversion of sucrose.

A hive bee evaporates water from the nectar by repeatedly regurgitating it from the honey sac and spreading it on the outstretched mouth parts. After this process, which takes about 20 minutes, the bee searches for empty cells and deposits the partly ripened nectar there by smearing it over the cell walls. In this position more water can evaporate than if all the nectar was placed in one cell.

During a very heavy nectar flow hive bees don't spend time evaporating the nectar, but simply place a small drop in each of several honey or brood cells hanging from the upper surfaces. When time permits, the nectar is processed as described above. Overnight the unripe honey smeared over the walls of many cells is further processed, and the water content reduced sufficiently for it to be packed into cells. Most cells are later sealed with a wax capping.

Gathering and storing water

Water is needed in the beehive mainly to dilute honey for feeding to larvae, and to cool or possibly humidify the hive interior. Water is collected in spring mostly to dilute brood food, and in summer to air-condition the hive. Bees don't readily collect from open bodies of water, but instead prefer to collect water from thin films, droplets or the edges of ponds (Fig. 4.16).

Fig. 4.16 Worker bee collecting water from a damp surface.

A water-gatherer dances vigorously on returning to the hive, sharing out its load among several nearby hive bees. It then grooms and feeds itself and returns to the field. Hive bees store water in specially constructed depressions in wax and propolis on frame top bars, and in some brood cells. Water is also stored in the honey sacs of resting worker bees, which act as reservoirs.

Water-carrying bees probably make 50–100 trips a day, with a load ranging from 25–50 mg each time. An average colony probably requires about 150 ml of water per day in spring for diluting larval food. Strong colonies in extremely hot conditions may need up to a litre of water a day.

Gathering and storing propolis

Propolis is a resinous material that bees usually collect from plant buds or scar tissue. Sometimes they will even collect putty or tar as a propolis substitute. Bees varnish the woodware in the hive with propolis, which is also used to cement hive parts together and plug cracks for weather-proofing.

Intruders such as bumble bees and mice may be stung to death, but are too big for the bees to remove. They are coated with propolis to embalm them and prevent their decay.

Fig. 4.17 Worker bee with propolis packed into its pollen baskets.

Propolis can be very sticky and its collection is a slow process. Bees tear propolis from the plant with their mandibles and pack it onto their pollen baskets (Fig. 4.17), taking up to an hour to gather a full load. They then return to the hive, where the propolis is removed by hive bees and put into place in the hive only as it is needed, a job that could take several hours. Bees may add wax to make the propolis more malleable. Once free of their loads, propolis gatherers fly from the hive to collect more.

RELATIVES OF THE HONEY BEE

Scientists have listed and described more than a million species of animals, and of these some 70 per cent are insects. The class of insects is divided into several groups: beetles, butterflies, bugs, flies and so on. One of these groups (or orders) is known as the Hymenoptera, which includes bees, ants, wasps and a few other types of insect. There are about 100,000 species of hymenopteran insects, some living separate or solitary lives, while others are social insects.

To many people the word 'bee' is synonymous with 'honey bee', but there are more than 20,000 different species of bees in the world, known collectively as the Apoidea. The only bees present in New Zealand are 33 species native to this country, and eight species that have been introduced since human settlement began.

Some authors spell 'honey bee' as one word, but it is a convention among biologists to spell two-part insect names as two words if the insect is what its name implies. For example, the names of true flies such as blow fly or hover fly are spelt as two words, but dragonfly and butterfly are not as they are not flies, just as silverfish is not a fish.

New Zealand native bees

We have used the term 'native' here rather broadly to include bee species that arose in New Zealand (endemic or indigenous), or arrived here without human intervention (adventive). New Zealand's native bees are mostly small black insects that nest in the ground. You can often see nest entrances on clay banks and exposed ground, and adult bees can be seen collecting pollen from flowers.

Twenty-eight of New Zealand's native bee species belong to the family Colletidae. They are solitary insects, with each female bee making a separate tube-shaped nest in the ground or in wood, in which she lays eggs in individual cells. The developing insects overwinter in their cells and emerge the following season, so there is no contact between generations.

Four other species belong to the family Halictidae. These bees have a similar sort of life cycle to the Colletidae except the female overwinters in the ground, and makes nests right throughout the warmer months. In one species, several females may forage from the same nest, indicating a primitive kind of social behaviour.

The recently discovered wool carder bee belongs to the family Megachilidae, and little is known so far about its life cycle in New Zealand.

Introduced bees

Apart from the honey bee, seven species of bees have been introduced to New Zealand, all for pollination purposes.

Alkali bee

The alkali bee, *Nomia melanderi* (Halictidae), is a ground-nesting bee native to the western United States of America that was introduced to New Zealand in 1964 because of its ability to pollinate lucerne. Its life cycle is similar to that of the soil-nesting native Colletidae, in that female bees fill individual cells with an egg and some food. The females are gregarious, preferring to nest close to other nests. In the US, large areas of ground can become filled with alkali bee nests.

Suitable nesting sites for the alkali bee are moist soils with high sodium or calcium salt levels. In New Zealand, nesting sites can be constructed by filling an excavated pit with appropriate soil, adjusting the salt balance and water table, and protecting it from rain and bird predation. Most of the alkali bee nests are in lucerne-growing areas of Marlborough, Canterbury and Central Otago. Alkali bees have not been extensively used, because of the work involved in preparing nest sites.

Lucerne leafcutter bee

Lucerne leafcutter bees are widely used in North America, where they have been introduced from western Asia to assist with pollinating lucerne. One species of leaf-cutter bee, *Megachile rotundata* (Megachilidae), was introduced into New Zealand from the US in 1971 for the same reason. Most leafcutter bees in the country are managed commercially for lucerne pollination. Populations are concentrated near lucerne paddocks in Marlborough (Fig. 4.18), with small numbers located in other lucerne-growing areas.

Leafcutter bee adults emerge from their cocoons in early spring, and after flying and mating, the female looks for a nest site in a hole about 6 mm in diameter and

Fig. 4.18 Lucerne leafcutter bee nests placed in a lucerne field for pollination.

100 mm long. She cuts oval pieces of leaf from nearby plants and makes a series of cells in the tube, laying an egg in each cell on a ball of nectar and pollen. The developing bee usually overwinters in the cells.

It is possible to manage leafcutter bees by providing straws, or drilled or grooved boards, as nesting sites. These are placed in shelters near lucerne paddocks and are used by female bees. After nesting is over, the cells can be removed and placed in cold storage. With careful timing and temperature regulation, the emergence of adult bees next season can be made to coincide with lucerne bloom.

Red clover mason bee

This bee, *Osmia coerulescens* (Megachilidae), was imported so it could be developed as a specialist, manageable pollinator for red clover seed crops. Numbers are currently small, and management methods are still being developed.

Bumble bees

The four species of bumble bee established in New Zealand are descendants of bumble bee queens imported from England between 1885 and 1906 to improve pollination of red clover. There are three long-tongued species — *Bombus hortorum*, *B. ruderatus* and *B. subterraneus*, and one short-tongued species, *B. terrestris*.

Bumble bees display some degree of social behaviour. Queens hibernate individually and start their own nests in spring, usually in old mouse nests or compost heaps. In each nest, bee numbers build up slowly in spring and more rapidly in summer, reaching a peak of perhaps several hundred adults. About three months after the nest is started, the population begins to decline. Eventually new queens leave the nest to hibernate, and the remaining bees die.

Bumble bees collect small amounts of nectar and pollen, which they store in small wax cups in the nest. They do not possess the advanced behavioural patterns of honey bees, such as division of labour among workers and a system of communication, but their longer tongues make them suitable for pollinating some plants.

In New Zealand some use has been made of bumble bees to pollinate crops such as red clover, blueberries under netting and greenhouse tomatoes. In 2009 British authorities made attempts to harvest *B. subterraneus* queen bees in New Zealand and reintroduce them to the United Kingdom, where the species was believed to have become extinct.

How to handle bees

The pressures of a strict schedule often force commercial beekeepers to open hives in all weathers, including wind and rain. If you are a hobbyist you can usually be more flexible, and should choose the most suitable time for working your bees, especially if the hives are in an urban environment. The best time is when the bees are flying and the weather is sunny, warm and not too windy.

The procedure for opening hives and examining colonies described in this chapter can be used when checking for colony strength, disease, food stores and brood production. It also forms a basis for specific hive manipulations such as varroa control, requeening, swarm control, and preparing hives for winter, which are described in later chapters.

Don't be reluctant to open your hives — with a gentle strain of bee, proper protective clothing and equipment, there is every reason to inspect your hives often and enjoy your beekeeping. Colonies are damaged more by neglect than by too frequent attention, especially now that varroa is widespread. However, be very gentle inspecting hives when there is little or no brood and no drones (in late autumn or winter). If you accidentally kill the queen at this time of year, the colony will most likely perish.

LIGHTING THE SMOKER

Before approaching the hive, check that the smoker is in good condition, with no holes in the bellows, and that the air vent and smoke chamber are free of excess tar and soot.

Next select some suitable fuel. The most commonly used smoker fuel is clean, dry, well-rotted sacking. New sacks are not as good, and should be placed outside to weather for a few months before use. Other suitable fuels include pine needles, non-tanalised wood shavings, straw, dry leaves, gum bark, lawsoniana hedge clippings, casuarina needles — or almost anything that will burn with a cool, white smoke. A base of old sacking with vegetation above it to keep the smoke cool is probably the best combination.

Light a small piece of fuel in the smoker, working the bellows very gently to make it flame. If the material is hard to light, you can use a small piece of paper to start the fire — but using too much paper creates difficulties later on, as the partially burnt paper restricts air flow in the smoker.

Add more pieces of fuel, continuing to puff the bellows. Rolls of sacking just smaller than the diameter of the smoker are particularly good, especially if they have some tinder or previously-burnt sacking in the centre of them. Keep these partly outside the smoker until they are well alight (Fig. 5.1), then put them in and close the lid. Some

beekeepers use the intense heat from small gas torches or burners to ignite the sacking, and the gas burners can also be useful for sterilising hive tools.

Dry pine needles make excellent fuel, and those who suffer from seasonal hay fever may find pine needles are less irritating than sacking. Place about 25 mm of needles in the bottom of the smoker. Then take a good handful of needles and light the bottom of this wad. Insert it into the smoker, but don't push it right down to the bottom. Work the bellows to get a good flame going, then gently push the wad of needles to the bottom of the smoker and work the bellows a few more times. Once the fire has taken hold and is

Fig. 5.1 Ensure the fuel is well alight before closing the smoker.

smouldering nicely with lots of cool smoke, add as much fuel as you can. This should be good for 10–15 minutes or more, and you can add more needles as necessary.

Be careful of sparks, especially if you are in a high fire-risk zone. Also take care if you are blowing on the wad of fuel to get it going, as sparks can burn the nylon gauze used in most veils.

If the smoker belches sparks or flames, add more fuel or put a wad of green grass inside on top of the fuel to cool the smoke as it comes out. Grass also helps keep pine needles and wood shavings from falling out of the smoker.

The smoker must produce cool, dense smoke to be effective. Using it correctly is an art, and success comes only with practice. There are three common causes of failure:

- The first is lighting a smoker and opening up a hive before checking that the smoker is going well. The smoker usually chooses to go out just as the bees become aggressive.
- Another beginners' fault is to have a poor fire with insufficient fuel, which needs to be pumped vigorously to keep it going. Smokers with too little fuel usually

shoot very hot air and little smoke onto the bees, burning their wings and making them aggressive rather than calming them.
- The third problem is not using the smoker enough to keep the bees under control at all times. The rule is to puff the smoker 'little but often'.

Once lit properly, a smoker with sacking fuel should keep going for at least 30 minutes without needing much attention. You should work the bellows occasionally to keep the fuel alight. Always keep some spare fuel in an overall pocket, so that the smoker can be refuelled as soon as it shows signs of going out. Stand the smoker on a hive when not in use for any length of time, as wind will keep the smoker burning. It may go out if put on its side on the ground.

Using a smoker while manipulating a hive can be awkward, especially when you need two hands for lifting boxes or removing frames. It can be annoying to have to keep putting the smoker down every time you want to perform another task in the hive. Because of this most beekeepers keep the smoker wedged between their knees when working hives, so it is always handy when needed. This is why smokers have a heat shield around the barrel.

WORKING A HIVE
Opening a hive
Put on all your protective clothing, making sure that there are no gaps for bees to get in. Approach the hive from the rear or side, and if at all possible avoid standing in front where the bees are flying. If the hives are in more than one row and face the same direction, work the front row first so that later you will work behind the disturbed colonies. Work a hive so that the sun is not shining into your eyes, which makes looking through a veil difficult.

Move slowly and carefully, as nervous or abrupt movements will make the guard bees react. Take reasonable care not to crush too many bees, as this can also make the colony aggressive.

First puff smoke gently into the hive entrance. This pacifies the guard bees, masks the alarm pheromones and helps to disorganise the colony's defence system. After a moment or two, lift off the lid, setting it upside down on the ground behind or beside the hive. It will later be used as a place to put supers or the second brood box, so should not be left in front of the hive where the stack of boxes will block the passage of returning foragers.

Use the hive tool to prise the inner cover off gently, then puff smoke across the top of the frames. In most cases you will want to look at both brood boxes, and you should start with the lower one first and replace the other boxes above the bottom one in order, examining each as you go. If you work from the top down, you will drive bees down into the next box. This will cause congestion there, and possibly result in bees hanging out of the entrance. You are also more likely to damage the queen if a box is very congested with bees.

To get to the lower brood box first, lift off the top one, place it in the upturned lid and

Fig. 5.2 'Cracking' the brood boxes.

Fig. 5.3 Removing brood frames for inspection.

cover it with the inner cover or hive mat. The two boxes may be difficult to separate or 'crack', and the easiest way is to use one hand to prise them apart with a hive tool while you lift the top box with the other hand (Fig. 5.2). Puff smoke into this gap before lifting off the top brood box.

If you are inspecting hives that have been neglected or just can't be opened easily, use a curved jemmy bar (available from most hardware stores), either alone or in conjunction with a hive tool to take some of the arm and wrist strain out of levering hive boxes apart. The leading edge of a larger jemmy bar should be ground or filed down to make inserting it between hive boxes a little easier.

A hive that has been badly neglected can be very difficult to open. If there is room to do so, lay the hive down on its back. Remove the floorboard and place it on the ground, then separate each box individually while keeping the smoker working. By laying the hive down you avoid having to bear the weight of the boxes while doing this. As each box is separated, you can place it back on the floorboard and rebuild the hive on-site.

Inspecting frames

To examine the frames in the bottom box, first puff smoke gently across the top bars. Use the hive tool to push all the frames in the box away from you (provided they are not Simplicity frames, which must be spaced by hand). Pull the nearest frame back towards you, breaking the propolis seal joining it to its neighbour. Remove this frame and examine it for the queen. The outside frame doesn't usually contain the queen (or brood), and providing she is not on it, set it on end against the opposite side of the hive near the entrance (Fig. 5.3). If she is on the frame, place the next frame outside the hive and return the one with the queen to the brood box.

With one frame out of the hive there is extra space for loosening and removing the frames in turn, so the risk of accidentally crushing the queen is reduced. Never pull frames directly out of the middle of the brood box. If there is a division board feeder in the hive, this can be removed first, so creating space to lever out the remaining frames.

Each frame can now be removed as you work in sequence across the box. Hold each frame above the hive while examining it in case the queen is on that frame and falls off (Fig. 5.4). The frames should be held by the lugs in the same orientation as they are in the hive. To look at the other side, turn the frame on end so that the top bar is vertical, rotate the frame through 180° on the axis of the top bar, and return the top bar to the horizontal (Figs 5.5–5.7). The frame will now be upside down with the opposite side of the comb facing you. This method of flipping the frame ensures that the comb's weight is always in the direction in which it has most strength. Reverse the process and replace the frame in the hive hard against the frame nearest to you. This stops the bees clustering together in wide gaps between the frames.

When the last frame has been inspected, push them all away from you across the box and replace the first frame in its original position. Smoke the bees off the

Fig. 5.4 Inspecting a frame with the beekeeper positioned so the sunlight comes over the shoulder.

Fig. 5.5 Examining one side of the frame, holding the bottom bar near you.

Fig. 5.6 Flipping the frame while holding the top bar vertical.

Fig. 5.7 Examining the other side, with the bottom bar away from you.

Fig. 5.8 Sting and venom sac in a beekeeper's finger.

tops of the frames before replacing the second brood box on the hive. (If you are looking for the queen, inspect the boxes separately.) The second box can then be inspected in the same way. When the inspection is complete, knock the bees off the inner cover onto the ground just in front of the hive entrance and replace the cover on the hive, followed by the lid.

During the inspection you will have to use the smoker from time to time. It is difficult to say exactly how often it should be used, as this varies from colony to colony and with different weather conditions. As a general rule, puff a little smoke across the top bars whenever the bees 'boil over' the tops of the frames. Try not to force smoke down between the frames and never smoke directly onto the face of a comb. The smoker is best used little but often, especially when looking for queens.

When you've finished with the smoker, block the nozzle with a tight wad of green grass inserted from the underside of the lid, or use a rubber bung or cork carried especially for the job. It may be necessary to block the bottom air inlet with grass as well. Never tip hot, spent fuel onto the ground because of the fire danger. Store the smoker inside a tin or in a metal box to minimise the chance of it setting fire to a vehicle or building. Laying it on its side in the metal box helps extinguish the fire.

COPING WITH STINGS

Wearing protective clothing greatly reduces your chances of being stung, but doesn't eliminate stings entirely. You can further minimise the chances of being stung by not using strong-smelling aftershave, perfume or hair spray before working with bees. Don't wear anything coloured blue, as bees see this colour particularly well, and don't wear an old leather watchstrap if you are handling bees without gloves.

When you are stung, don't panic or you will probably receive a lot more stings. Certainly don't start swatting at bees, as your rapid hand movements will only make things worse. The first priority is to remove the sting, as this will reduce the amount of venom being pumped into your body. Scrape the sting out with your hive tool or fingernail, being very careful not to squeeze the poison sac (Fig. 5.8) and thus drive more venom into the skin.

The odour of the sting attracts other bees to sting in the same spot, so smoke the affected area heavily to disguise this smell.

If you are stung under the veil, resist the very strong urge to take it off to deal with the sting. Either reach carefully under the veil and scrape out the sting, or move away to a building or vehicle to remove your veil. Any stray bees following you will be attracted to the windows and will be less likely to sting. Retreating into a thicket of scrub usually helps you to lose a lot of the pursuing bees.

Fig. 5.9 Stings left in the leg of a pair of overalls after a bad stinging attack.

Sometimes a colony gets out of control and you may get badly stung. If this happens, retire from the hive and make sure that your protective clothing is securely in place. Smoke your gloves or wash them to disguise the odour of stings, return to the hive and gently close it up. Don't hurry over this. Quick, jerky movements will only make the bees more aggressive, and no matter how fast you move you still can't beat an angry bee. After a bad stinging (Fig. 5.9), all protective clothing should be washed to remove sting odours and alarm pheromones. Ask a doctor to suggest which antihistamines should be carried in case you — or a bystander — experience an abnormal reaction.

Knowing the fundamentals of handling bees is the foundation for all of the colony management techniques described in later chapters. For beekeeping to be enjoyable you must be confident and relaxed when working hives, which will come with practice.

6 Nectar and pollen sources

Many people start beekeeping because of their interest in the environment and the relationship between insects and flowers, and in particular the importance of bees in pollination. If you don't know much about flowering plants when you start beekeeping you soon will, or at least should. Observing which plants flower in your area and when, which flowers are visited by bees, and what bees take from the flowers, is an important part of being a competent beekeeper. It can also be interesting trying to deduce what the bees have been working on based on the colour of their pollen loads, especially in cities.

Nectar and pollen provide bees with all of their nutritional requirements. Nectar contains sugars as an energy source, and pollen supplies protein, fats, vitamins, minerals and other necessary substances.

When nectar and pollen are not available outside the hive, bees rely on the stores in their combs or on supplementary feed provided by the beekeeper. Sugars can be supplied relatively easily as cane sugar or honey, but protein is more difficult to provide and pollen shortages can be a real problem in some areas.

Food sources that flower in spring are particularly valuable to honey bees, because colonies are then building up in population from low winter levels to a peak before the main honey flow (more properly known as the nectar flow), which starts in mid-November to mid-December in most areas.

For their nectar and pollen needs, bees forage on cultivated crop or pasture plants, native trees and shrubs, and introduced weeds. In many areas, land development has reduced the diversity of vegetation. This has resulted in a shortage of bee forage, particularly during spring, so beekeepers must shift apiaries to good spring areas or provide supplementary feed. The need for both practices can be partly offset by planting early spring-flowering nectar and pollen sources.

It is seldom possible to plant large areas of land in plants solely for the purpose of bee feed, but many plantings for other purposes can also benefit bees. Shelter and erosion-control trees, garden and park amenity plantings, and roadside plantings can all be useful

Fig. 6.1 Apiary in an area of viper's bugloss, central South Island.

sources of nectar and pollen. Before planting, ensure that species are not governed by your regional council's pest management strategy or the national pest plant accord.

PLANTS IMPORTANT TO BEEKEEPERS
Tables 6.1 to 6.4 list some of the common plant species important to beekeepers in New Zealand. Many hundreds more trees and shrubs can be planted as nectar and pollen sources, and information about some of these is given in the books listed at the end of this book.

Table 6.1 Nectar and pollen sources in winter and early spring
(June to September)

Key to nectar and pollen values

N>P	used by bees more for nectar than for pollen
P>N	used more for pollen than for nectar
N=P	equally valuable for nectar and pollen
P	pollen source only
S	bees may gather surplus honey from this source in favourable conditions

Name	Final height (m)	Nectar and pollen value	Surplus honey production	Comments
TREES				
Acacia baileyana Cootamundra wattle	10	P	—	Fast-growing, easily cultivated from seed. Ornamental, half-hardy.
Banksia integrifolia banksia	6	N	—	Hardy coastal tree that requires free-draining soil. Found in the Auckland region.
Hakea acicularis spiny hakea	3	N=P	—	Shelter plant, growing wild in the Auckland region and Golden Bay.
Prunus persica peach, nectarine	6	P>N	—	
Prunus species ornamental cherry	5	P>N	—	Ornamental, deciduous. Easily established from cuttings or seedlings.
Pseudopanax arboreus five finger	8	N>P	—	Native, common in regenerating bush. Flowers July–September. Found throughout New Zealand up to 750 m.
Salix caprea pussy willow	12	P>N	—	Early source, important for pollen particularly in areas of pollen shortage.
Salix fragilis common willow	15	N=P	—	Flowers September–October, valuable source of nectar and pollen. Common near swamps and waterways, where it may block water flow.
Salix matsudana Peking willow, matsudana willow	5	N=P	S	Suitable for shelter and erosion control. Very fast-growing; grown from cuttings or poles.
Sophora tetraptera, *Sophora microphylla* kowhai	10–12	N=P	—	Native plants of bush margins and remnants, often cultivated for ornamental value. Copious nectar August—October. May be narcotic to bees in some areas and some seasons.

Name	Final height (m)	Nectar and pollen value	Surplus honey production	Comments
SHRUBS				
Chamaecytisus palmensis tree lucerne, tagasaste	5	N>P	—	Rapidly grown shelter plant, good value for bees. Susceptible to silver leaf and stem borers.
Ribes sanguinem flowering currant	3–5	N=P	—	Ornamental, common in a few localised areas. Shade-tolerant.
Rosemarinus species rosemary	2	N=P	—	Common culinary herb and ornamental. Prostrate form useful for stabilising banks.
Ulex europaeus Gorse	3	P	—	Excellent pollen source and erosion control plant, but an invasive species that may be controlled as a pest plant by some regional councils.
HERBS				
Erica lusitanica Spanish heath	0.5	N>P	—	Excellent source for maintaining colonies during winter and spring. Begins flowering in April in some localities. Common on poor soils and roadside cuttings.
VINES				
Rubus species bush lawyer	—	N=P	—	Native, common in bush areas.

Fig. 6.2 White clover, one of New Zealand's premier honey sources.

Table 6.2   Nectar sources in spring and early summer
(October and November)

Key to nectar and pollen values
N>P   used by bees more for nectar than for pollen
P>N   used more for pollen than for nectar
N=P   equally valuable for nectar and pollen
P   pollen source only
S   bees may gather surplus honey from this source in favourable conditions

Name	Final height (m)	Nectar and pollen value	Surplus honey production	Comments
TREES				
Acer pseudo-platanus sycamore	20	N>P	—	Shelter or specimen tree, deciduous.
Aristotelia serrata wineberry, makomako	10	N>P	—	A minor source. Native, found in regenerating bush. Erosion control species.
Cordyline australis cabbage tree	20	N>P	S	A native plant common in pastureland and scrub, especially wetter areas. Also an ornamental. Profuse nectar secretion October to early December.
Corynocarpus laevigatus karaka	15	N=P	—	Native tree of coastal areas. Nectar very toxic to bees.
Elaeocarpus dentatus hinau	18	N>P	S	Beautiful native tree with heavy flowering November–December. Found throughout New Zealand to 600 m.
Fuchsia excorticata kotukutuku, konini	5–14	N=P	—	Deciduous native, found in bush and regenerating cover. Good source of nectar and conspicuous deep blue pollen from August–December.
Knightia excelsa rewarewa	30	N=P	S	Pioneer species of native bush. Not reliable, but yields heavily on occasions. Found in the North Island and Marlborough Sounds. Decorative timber, dense and coarse-grained. Amber honey.
Malus sylvestris apple	15	P>N	—	Moderate quantities of pollen and nectar gathered if weather is fine.
Pittosporum crassifolium karo	9	N=P	—	Native tree also found as a garden ornamental.
Pittosporum eugenioides lemonwood, tarata	12	N=P	—	Native tree, useful for shelter or hedge. Found throughout New Zealand.
Pittosporum tenuifolium kohuhu	9	N>P	—	Native, found throughout New Zealand except West Coast of the South Island. Common hedge plant.

Name	Final height (m)	Nectar and pollen value	Surplus honey production	Comments
Pyrus species pear	17	P	—	Attractive pollen, but nectar usually too low in sugar concentration to attract bees.
Quintinia acutifolia Westland quintinia	12	N>P	S	Similar to kamahi in appearance. Flowers October–December. Found throughout New Zealand; local in the North Island but common on the West Coast of the South Island.
Robinia pseudoacacia false wattle, black locust	10	N=P	—	Deciduous legume. Not all varieties give good nectar secretion. Rapid growth, easily established from cuttings. Suckering can be troublesome. Timber useful for fence posts and firewood.
Weinmannia racemosa kamahi	25	N>P	S	Native tree of regenerating and virgin bush. Flowers November–January. Reliable producer of distinctively flavoured honey.
Weinmannia sylvicola tawhero, towai	20	N>P	S	Flowering similar to that of kamahi. Found to latitude 38° south, particularly on forest margins.

SHRUBS

Name	Final height (m)	Nectar and pollen value	Surplus honey production	Comments
Berberis species barberry	4	N>P	S	Excellent source, flowers September–November. Common shelter species.
Buddleia salvifolia buddleia	4	N>P	—	Salt and wind tolerant. With regular trimming will make good hedge plant. Fast-growing. Second flowering in late summer.
Crataegus oxyacantha hawthorn	5	N=P	S	Shelter plant, excellent for bees.
Cytisus scoparius broom	3	P	—	Good pollen source in September–November. Spreads in waste areas.
Hebe species koromiko	1	N=P	—	Over 100 species native to New Zealand. Many lowland forms good for bees. Pioneer species in natural ecosystems, often grown as ornamentals.
Phormium tenax flax	4	N=P	S	Common throughout New Zealand, especially in wetter areas. Honey dark and strong.

HERBS

Name	Final height (m)	Nectar and pollen value	Surplus honey production	Comments
Cyathodes fraseri patotara, bronze heath	—	N=P	—	Native heath-like plant. Excellent spring source.
Ranunculus species buttercup	—	N>P	S	Found in pastures, particularly in wetter areas. Surplus honey is dark and of medium flavour.

Name	Final height (m)	Nectar and pollen value	Surplus honey production	Comments
Taraxacum officinale dandelion	—	N=P	—	Common herb in pasture areas. Good for spring build-up, also can produce a good flow in autumn.
Thymus vulgaris thyme	0.5	N>P	S	Cultivated as culinary herb, grows wild in parts of Central Otago. Honey very strong in flavour. Flowers mid-October to November.

VINES

Name	Final height (m)	Nectar and pollen value	Surplus honey production	Comments
Actinidia deliciosa kiwifruit	—	P	—	Gold variety flowers in October and green variety in November–December. Source of cream pollen.
Rubus species berry fruit	—	N>P	S	Raspberry and blackberry are particularly valuable as nectar sources, occasionally yielding surplus honey. May produce honeydew.

CROPS

Name	Final height (m)	Nectar and pollen value	Surplus honey production	Comments
Brassica species brassicas	—	N>P	S	Includes chou moellier, canola and turnip. Copious quantities of nectar; honey white, mild and fast-granulating.

Fig. 6.3 (inset) Pohutukawa, a magnificent nectar source in northern coastal areas.
Fig. 6.4 Broom, one of many species that are valuable forage plants for bees.

Table 6.3   Nectar and pollen sources in summer
(December–March)

Key to nectar and pollen values
N>P   used by bees more for nectar than for pollen
P>N   used more for pollen than for nectar
N=P   equally valuable for nectar and pollen
P   pollen source only
S   bees may gather surplus honey from this source in favourable conditions

Name	Final height (m)	Nectar and pollen value	Surplus honey production	Comments
TREES				
Eucalyptus ficifolia red gum	15	N>P	S	Showy ornamental, vivid red flowers late January–February. Surplus honey if sufficient number of trees. Young plants frost-tender, older ones relatively hardy.
Eucalyptus viminalis manna gum	25	N>P	S	Heavy flowering in February–April, keenly worked by bees.
Melicytus ramiflorus mahoe	10	N=P	—	Native tree found throughout New Zealand. Yields dark amber nectar and cream pollen November–December.
Metrosideros excelsa pohutukawa	20	N>P	S	Coastal native tree, salt-tolerant. Flowers December–January, profuse quantities of nectar. Honey white and fast-granulating. Closely related to *Metrosideros kermadecensis,* a common hedge plant.
Metrosideros robusta rata	25	N>P	S	Native forest tree, throughout North Island and down to about Greymouth. May hybridise with southern rata. Flowering more reliable but less intense than that species.
Metrosideros umbellata southern rata	20	N>P	S	Found throughout New Zealand but local in the North Island. Common from Greymouth south. Flowers December–April, depending on altitude. Heavy flowerings occur irregularly every few years. Extremely good nectar source. Both rata species produce white honey with a distinctive but delicate flavour.
Ixerba brexioides tawari	17	N>P	S	Native tree of virgin bush above about 300 m, to 38° south. Magnificent flowering from November to January and copious quantities of nectar. Honey white, very mild and sweet in flavour. May be prone to fermentation.

Name	Final height (m)	Nectar and pollen value	Surplus honey production	Comments
Kunzea ericoides kanuka	15	N>P	S	Light amber honey with flavour and aroma similar to, but more delicate than, that from manuka. Partly thixotropic (jelly-like) and rapidly granulating.
Leptospermum scoparium manuka	4	N>P	S	Pioneer species, especially on infertile soils. Honey amber and strong-flavoured, thixotropic and difficult to extract from combs. Honey from some areas has significant antibacterial properties.
Ligustrum chinensis Chinese privet	3	N>P	—	Flowers in the spring and autumn. Pale, aromatic bitter honey sometimes produced.
Schinus molle Peruvian pepper tree	15	N>P	—	Ornamental. Half-hardy.
Tilia species lime, linden	20	N>P	S	Specimen tree in urban parks and streets. Short, intense nectar flow in December. Timber valued for carving and furniture making.

SHRUBS

Name	Final height (m)	Nectar and pollen value	Surplus honey production	Comments
Lycium horridum boxthorn	4	N>P	S	Good shelter plant, will withstand salt winds. Common in Taranaki. Very thorny, needs trimming. Good late nectar source.

HERBS

Name	Final height (m)	Nectar and pollen value	Surplus honey production	Comments
Astelia species astelia	—	N>P	S	Mostly epiphytes in native bush. Objectionable honey produced spring to autumn depending on species.
Calluna vulgaris ling leather	—	N=P	S	Confined to central volcanic plateau of North Island. Honey thixotropic (jelly-like) and strong-flavoured, and may have high moisture content.
Cirsium species, *Carduus* species, *Silybum* species thistles	—	N>P	S	Includes Scotch, nodding, winged and Californian thistles. White honey produced December–March depending on species.
Echium vulgare viper's bugloss (often called blue borage, though it is not a borage species)	1	N>P	S	Abundant in drier areas of the South Island and parts of Hawke's Bay. Dull white, delicate-flavoured, slow-granulating honey.
Foeniculum vulgare fennel	2	N>P	S	Flowers February–March. Unpleasant-tasting honey, but good for winter stores.

Name	Final height (m)	Nectar and pollen value	Surplus honey production	Comments
Hypochoeris radicata catsear	—	N=P	S	Also capeweed, hawksbeard, and other related species. Yellow cappings. Good for autumn stores.
Lotus species lotus	—	N>P	S	Found in pasture lands and roadside banks, particularly damp areas. All species useful. Honey mild in flavour and yellowish in colour.
Medicago sativa lucerne	1	N>P	S	Good nectar source in dry areas such as Marlborough, Canterbury and Central Otago. Water-white honey similar to that of clover.
Melilotus species sweet clover	—	N>P	S	Found principally in dry areas such as river flats in the South Island.
Mentha pulegium penny royal	—	N>P	S	Creeping weed of pasture and waste land. Flowers January–March, surplus honey produced only in long dry spells. Honey strongly aromatic, may be prone to fermentation.
Trifolium hybridum alsike clover	—	N>P	S	More resistant to frosts than white clover, used in over-sowing mix in high country. Excellent nectar source, honey similar to that from white clover.
Trifolium repens white clover	—	N>P	S	One of New Zealand's most important honey plants, along with manuka. Yields nectar November–February, depending on location. White, delicately flavoured honey.
Trifolium pratense red clover	—	N>P	S	Sensitive to soil conditions, may be erratic in nectar yield. Honey similar to that from white clover.

Fig. 6.5 Manuka, source of a honey valued for its antibacterial properties.

Fig. 6.6 Ling heather.

Table 6.4   Nectar and pollen sources in autumn and early winter
            (April–May)

Key to nectar and pollen values
N>P   used by bees more for nectar than for pollen
P>N   used more for pollen than for nectar
N=P   equally valuable for nectar and pollen
 P   pollen source only
 S   bees may gather surplus honey from this source in favourable conditions

Name	Final height (m)	Nectar and pollen value	Surplus honey production	Comments
TREES				
Eucalyptus leucoxylon rosea winter-flowering pink gum	15	N=P	—	Hardy, withstands winds and medium frosts. Ornamental, good nectar source.
Hoheria populnea lacebark, houhere	11	N=P	—	Native found naturally to 38° south, but now cultivated throughout New Zealand as an ornamental. Profuse flowering.
VINES				
Metrosideros species rata vine	—	N=P	S	Good winter stores or surplus in some years.
CROPS				
Helianthus annuus sunflower	2+	N=P	S	Mild amber honey and yellow-orange pollen.

Fig. 6.7 Roadside plantings such as this bank of lotus can be useful bee forage.

Colony management 7

In this chapter we give a general picture of what you as a beekeeper will have to do to manage your hives through a complete season. Specific management techniques are described in the following five chapters.

To manage colonies well you need to understand honey bee biology (Chapter 4), and adapt the colony's natural patterns to suit your aims as a beekeeper.

In nature, honey bee colonies follow a particular seasonal pattern in temperate climates such as New Zealand's. They start from low numbers of worker bees in winter, and increase in size during spring when they often multiply and disperse by swarming.

Then comes a period called the 'honey flow', when the colony stores enough surplus honey to survive the winter. In reality it is a nectar flow, but we use the term honey flow to reflect beekeepers' common usage.

Hive populations decrease in autumn and dwindle slowly over winter. In many warmer areas, and often in cities with varied nectar and pollen sources, hives with young queens can continue to rear brood throughout the winter.

Whether you want to let your colonies follow this natural cycle or whether you want to adapt it depends on your objectives in beekeeping. If you do want to maximise honey production from your hives, this natural cycle must be modified. While not all hobby beekeepers have the goal of maximum honey production, this chapter is aimed at those who do want to get the most from their bees. It can also apply to beekeepers who want to provide hives for pollination.

The aim of beekeeping for honey production is to manage colonies in such a way that the maximum population of foraging workers coincides with the start of the main honey flows of the area. Beekeepers do this by:

* finding out when the main honey flows occur
* managing colonies so they build up high numbers of foragers *before* these honey flows start
* managing hives after the honey flow so they overwinter well, and can build up to

be in good condition again for the next honey-flow period
- managing varroa where it is present.

The importance of high bee numbers at the honey flow cannot be overstressed. Strong colonies store a disproportionately large honey surplus compared with smaller colonies. In other words, a certain number of bees divided into several small colonies will gather less honey than the same number of bees working in fewer large colonies. Efficiencies of scale in larger colonies mean that a higher proportion of bees can be free of hive duties and available for foraging.

Having strong colonies for the honey flow is important, but correct timing of the colony build-up is vital. Honey bees take three weeks to develop from egg to adult, and about a further three weeks before the adults start to forage. Put these two facts together and you can see that a rapid increase in egg-laying won't produce any change in forager force for approximately six weeks.

All too often, beekeepers allow their colonies to build up on the honey flow rather than building them up for the flow. The difference is often mystifying to a beginner. Imagine adding honey supers to two adjacent hives. At the end of the honey flow one hive has filled two boxes with honey, the other less than one box. The hives seem equally populous. What went wrong?

Fig. 7.1 shows the difference between hives that build up on the flow, and those that are built up for the flow. Colony A has been built up in advance of the honey flow, the beekeeper having allowed for the time lag between egg-laying and foraging. Colony B has built up during the flow. At the end of the honey-gathering period both colonies are the same strength. But one will have produced much more surplus honey than the other.

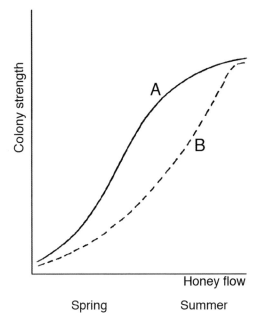

Fig. 7.1 Preparation of hives for the honey flow: (A) a colony that builds up for the flow; (B) a colony that builds up on the flow.

The difference between these hypothetical colonies is the difference between a well-managed and productive hive, and a poorly managed and unproductive one. Most beekeepers want productive hives, and the process of managing the colony to achieve this is outlined below and in the next few chapters.

SPRING

As daytime temperatures increase above 14 °C during spring, honey bees in a colony stop clustering during daylight hours and begin to fly more and more. They begin to gather nectar, pollen and water, and defecate on the wing during cleansing flights. Brood rearing recommences or speeds up, and the colony population begins to increase. The Carniolan strain of bee may not begin this build-up as early as the Italian strain.

Early spring management

You can examine your colonies in spring when the bees are flying and have stopped clustering. In most districts a two-storey hive that has been well prepared for winter should not need to be opened before mid-September. Hives wintered as singles or nucs will need to be inspected in August. Some beekeepers carry out their early spring checks simply by lifting the back of the hive to feel its weight — only particularly light hives need to be examined and fed.

Fig. 7.2 Inspecting hives in spring.

Choose a fine, warm afternoon for your first spring check (Fig. 7.2). Examine the brood for American foulbrood before stripping the hive right down to the floorboard and scraping off the wax and other debris that will have accumulated over winter. It is good hygiene to do this at least a few metres away from the hive site. Depending on what varroa treatment you used in the autumn, what time in September you are making this inspection, and when the surplus honey flow starts in your area, you may need to consider applying your first varroa treatment at this time.

Remove any damaged hive parts. Mouldy frames don't need to be removed, as the bees will clean them up as the colony gets stronger. If you replace any damaged combs in the bottom brood box, do this with drawn comb rather than foundation. One or two frames of foundation or waxed plastic frames can be placed in the second brood chamber, but only if there is a honey flow on or you are feeding heavily with sugar syrup.

Replace the bottom brood box on the hive stand and check through it. In a normal double-brood-chamber hive very few bees or stores will be in the bottom box, except in city areas where brood rearing often starts earlier than in rural districts. It is normal to swap the two brood boxes at this point, as part of your swarm-prevention measures (see Chapter 9).

Carefully examine the frames in each box in turn. Look at the brood (Fig. 7.3) to see if the colony has a laying queen (is 'queenright'). If there is no brood, and especially no eggs, the colony is probably queenless and should be united with another colony or nucleus. If there are a lot of domed cappings of drone brood in worker cells, and multiple eggs in cells, then

Fig. 7.3 A solid frame of brood in a colony building up well in the spring.

the queen has probably turned into a drone layer or there are laying workers present. As spring is usually too early to purchase queens, the queen should be killed and the colony united with another one.

It is very important to check for food stores. The colony should never have less than the equivalent of three frames of honey. If there is less honey in the hive, start feeding immediately.

The position of honey frames in the hive is also important. The colony will be clustering in the middle of the brood box, with the honey on the outside. If there are several empty frames between the cluster of bees and the honey frames, shift the honey alongside the cluster. But take care never to put the honey frames into where the bees are clustering, as this will split the cluster into two smaller ones. Bees do not readily cluster on honey and the brood may get chilled.

If there is a lot of moisture on the inner cover or on the hive walls, raise the inner cover on matchsticks to allow moist air to escape.

While at the apiary, check that your registered number is displayed, and the fencing is secure. Control any weed growth around the hives. Scrub cutters and domestic weed clippers do the job, although regular use is necessary. An easy solution is to use a long-lasting herbicide in spray or granular form around the hives. Generally herbicides are not toxic to bees, although you should take care not to spray directly into hive entrances as some wetting agents used in herbicides can be toxic to bees. Weeds are also kept down if hives are put on sheets of roofing iron, but this may make it difficult to work the hives if you need to stand on the roofing iron to do so. Some beekeepers use weed mats or old pieces of carpet or vinyl to suppress weed growth.

There is no hurry to take off entrance reducers, which can be left on until a later visit in case of continued wasp or rodent activity.

Robbing can be a problem in spring, so take care not to expose honey for longer than necessary and try not to spill any sugar syrup when feeding hives. If robbing does start, close all hives and reduce the entrances of any that are weak or are being badly attacked.

During spring, bee colonies require a lot of water for diluting honey being fed to larvae. Most areas in New Zealand have adequate water for bees, but in unusually dry areas you should provide extra.

Management up to the honey flow

September, October, and November are the busiest months of any beekeeper's calendar, and this busy spring period extends into December for those doing pollination. Requeening is often carried out at this time, and the hives must be built up at the right pace for the area.

You should check your hives at least every three weeks during this period. Carry out proper swarm prevention measures. Check the colony for feed, and add honey or sugar if stores are insufficient. Examine the brood for signs of diseases, and apply varroa treatments. Entrance reducers can be taken off the hives in mid-October.

Oddly enough, it is just before the surplus honey flow — mid-December in many districts — that hives can die of starvation. November is a critical month as colonies are very populous and consume food stores quickly. There is often a gap in the flowering of nectar sources, leading to a dearth of incoming nectar. Never let colony stores dwindle below the equivalent of three frames of honey. Feed the bees if stores drop to this point, and check hives weekly if the weather has been bad.

The essence of correct spring management is timing. If you are two days late with a feed of sugar syrup a hive might die. If the whole build-up is too slow the colony will miss the beginning of the honey flow, and build up on the flow instead as described earlier. A colony that builds up too early may swarm, or even if it does not swarm, will peak in population well before the beginning of the flow and may require a lot of feeding. To prevent this happening it may pay to split strong hives during the build-up and reunite them as the honey flow begins. This can act as a combined swarm prevention and re-queening method. Techniques are described in Chapter 10.

HONEY-FLOW MANAGEMENT

Correct hive management during spring will mean that colonies have built up without swarming, and at the beginning of the honey flow will have a large population of workers old enough to forage for nectar. Any colony that does not fully occupy more than one brood box at the beginning of the honey flow should be united with another.

Beginning of the flow

It is easy to identify the beginning of the honey flow. Foraging activity increases and the colony is occupied with gathering nectar, so is less aggressive to the beekeeper and no robbing takes place. Inside the hive you will see fresh white wax being secreted to build up the honey cells, and some of this is placed on the top bars of frames. The bees are said to be 'white waxing' (Fig. 7.4). Fresh nectar is also conspicuous in the combs. If you tip a frame on its side and can shake out a shower of nectar, then the nectar was collected that day.

Fig. 7.4 A strong colony building fresh burr comb on the top bars, or 'white waxing'.

Supering up

Before adding honey supers to a hive, ensure that all varroa treatments have been completed, and any treatment strips removed to minimise the risk of residues contaminating wax, propolis or honey.

The timing of adding honey supers to a hive is critical. Adding supers relieves congestion in the hive and helps reduce swarming. Honey supers contain space needed for ripening and storing honey. Incoming nectar contains 50–80 per cent water, and so requires more room while the bees are processing it than when it is stored in concentrated form as honey with less than 18 per cent water content.

If not enough supers are put on a hive the bees will store honey in the brood nest. This restricts the queen's laying, and the workers forage less for nectar. Even if the hive is then supered up, the bees seem to take some time to regain lost momentum, and several days

Fig. 7.5 A strong colony in need of another super, 'hanging out' of the hive.

of potential honey-gathering may have been lost. When your main honey flow is only a few weeks long, that can be significant.

The first honey super will often have to be put on before the main honey flow starts, particularly if there are significant early honey flows. Putting it on early (mid to late October in most regions) also helps to prevent swarming, and may assist with controlling wax moth damage in the stored combs.

The first super should contain drawn combs if possible, as foundation does not give the bees any more room until they can build cells on it. Strong colonies, high temperatures and a good honey flow are needed for them to build comb on wax or plastic foundation. Any foundation drawn out on an erratic honey flow will not be drawn out well, and will not be attached to the bottom bar. Bees may even chew the wax foundation and make holes in it. Plastic frames need a strong honey flow to be drawn out and filled, whether they are waxed or not.

Additional supers should be added before the bees desperately need them (Fig. 7.5). This maintains the momentum of the colony, and the addition of plenty of space seems to stimulate the bees to gather more nectar. An old idea about supering a hive was that you should wait until the bees were white-waxing the top bars of a box before adding another super. In fact, the white-waxing stage is really the latest that a super should be added, and supers should be put on earlier than this during a honey flow.

Supers are usually added on top of the previous super, in a process called top supering. This is the easiest way to put on supers, as less lifting is involved and you can easily check the super to see when another might be needed.

Bottom supering involves lifting off one or more nearly full boxes and adding the new super directly above the brood nest. It demands a lot more work, which is warranted only if supering has been delayed and combs in the top super have been completely capped over, or you believe the honey flow is nearly finished. It often pays to under-super boxes of comb honey as bees draw foundation better nearer the brood nest.

It is impossible to give firm guidelines about how often to super a hive, as honey flows vary greatly. You should obviously put on enough supers to last the hive until your next visit. This may be one, two, or even more supers, depending on the flow. On a good honey flow, strong hives will fill a super in one or two weeks. Supers can be filled in as little as two days, but only on very strong colonies in an extraordinarily good honey flow.

If in doubt about how many boxes to put on, err on the generous side as bees can never fill honey supers that are still in storage. Hives don't suffer from having too much room during the honey flow. All that happens if the honey flow stops before the supers are filled is that some rearrangement of frames will be necessary to give the beekeeper full boxes of honey.

Fig. 7.6 Wax foundation starting to be drawn out.

Using wax or plastic foundation in honey supers

As well as the traditional wax foundation (Fig. 7.6), beekeepers use plastic frames that include foundation, and plastic foundation sheets inserted into wooden frames. The plastic foundation sheets (Fig. 7.7) should be waxed before they are put into a hive — they can either be purchased pre-waxed or the beekeeper can wax the sheets. In the discussion that follows, the word foundation will be used to mean either wax sheets or waxed plastic foundation.

Foundation is drawn out properly only on a good honey flow or if you feed a lot of sugar syrup, so don't use foundation at other times unless you have no drawn combs.

Unless a honey flow is particularly good, supers of foundation should be 'baited'

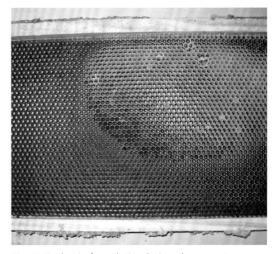

Fig. 7.7 Plastic foundation being drawn out.

when they are put on a hive. Pull out two partly-capped honey frames from the top super already on the hive and place them near the centre of the box of foundation, separated by a couple of new frames. Put the two frames of foundation displaced from the new super down into the centre of the top super to make up the spacing. Then add the new super of foundation. Baiting the super of foundation in this way stimulates the bees to move up more quickly to start drawing out comb, and helps to keep up the momentum of the colony.

Honey supers normally contain eight Hoffman frames evenly spaced by hand, or Manley frames that are self-spacing at eight per box. Use nine or even 10 Hoffman frames per super if foundation is used, to ensure they are evenly drawn out and to minimise the bees' tendency to fill large gaps with their own comb. Frames are later spaced at eight per box to ensure the bees build fat combs, which makes uncapping much easier.

Fig. 7.8 Queen excluder in use on a hive.

Queen excluders

Opinions vary on the use of queen excluders (Fig. 7.8) during the honey flow. The advantages of using them are:

- it is easier to find the queen, as she is confined to boxes below the excluder
- brood is also confined to the brood boxes, where it is more easily inspected for American foulbrood before the honey is taken for extraction
- fewer varroa strips may be needed, as label instructions for some miticides specify a number of strips per brood box, and the brood is confined
- harvesting honey is easier, especially if using escape boards, as bees will not readily leave brood frames
- honey frames are kept white, and free of the debris of brood-rearing
- less pollen is stored in the honey supers.

The disadvantages of queen excluders are:

- they may slow down bee movement into the honey supers
- if not used correctly they may cause congestion in the brood nest
- they will restrict ventilation of the hive if allowed to become clogged with honey and wax
- they may become blocked with drones if the queen is inadvertently trapped above the excluder
- they are expensive and are easily damaged.

If working a hive with a queen excluder on it, be careful not to trap the queen in the honey supers. Open the hive in the normal way, and set the honey supers in the upturned lid. Smoke the queen excluder before removing it from the top of the brood nest, turn it over and look for the queen. If you can't see her, put the excluder squarely on the top of the honey supers in the upside-down position. When reassembling the hive, flip the excluder over to put it back on the brood nest in its original orientation. There is little chance of the queen ending up in the supers if you follow this routine.

Never insert a hive tool between the wires of an excluder to lever it off a box, as a queen will readily pass through an excluder if the wires are bent even slightly. When removing honey supers insert your hive tool above the excluder rim and lever down. This applies more pressure to the bottom edge of the super above than to the excluder. To check an excluder for damage, hold it parallel with the ground at eye level and look along the wires.

If queen excluders are not used, wait until the bees make a ring of honey on the tops of the frames in the second brood box before adding a honey super. Another useful technique is to rearrange the brood nest when putting supers on — placing the queen and most of the brood in the bottom brood box, with sealed honey and sealed brood in the second box. The honey acts to some extent as a natural queen excluder.

Queens are a little less likely to go up into three-quarter depth boxes than full-depth, and less likely to lay in Manley frames than in the more closely spaced Hoffman frames, although they will still do either at times.

MANAGING COLONIES FOR POLLINATION SERVICES

The best colony for pollination is one that has both a high population of foraging bees and a high demand for pollen, when the target crop flowers. Based on the principles of colony management set out in this chapter, and the bee biology outlined in Chapter 4, in practice this means:

- Workers start foraging at about three weeks of age, so the colony population should be high three weeks before the crop starts flowering.
- Colonies are stimulated to collect pollen by the presence of unsealed brood, so good pollinating units must have a young queen that has a high egg-laying rate. To achieve this for early-flowering crops, you may need to stimulate the colony with supplementary feed in the build-up to the pollination period.
- For late-flowering crops, frames of honey and/or brood may need to be removed to give the queen adequate room to lay, or to try to prevent swarming.

Fig. 7.9 Hives being used to pollinate berryfruit.

- Hives in commercial kiwifruit orchards, especially the green variety, benefit from being fed 1–2 litres of sugar syrup every few days, or three or four times while the hives are in the orchard, to stimulate pollen collection.
- Pollination colonies must have plenty of good-quality comb space available, otherwise brood-rearing will slow down or the colony may even swarm.

Chapter 19 covers pollination hive management in more detail.

AUTUMN

When working hives in late summer or autumn be particularly careful to avoid robbing. Leave hives open for as short a time as possible, and never drop scraps of comb in the apiary. Be particularly careful not to spill sugar syrup when feeding.

Late summer can be a critical time to apply varroa treatment. This is especially so in the

Fig. 7.10 Inspecting the brood nest in autumn before wintering a hive down.

acute phase of infestation — the first two to four years after varroa arrives in your area — when untreated and feral hives are collapsing and absconding, and invasion of your hives by varroa is at a peak. Bee-keepers may need to harvest honey earlier than they did in the past in order to apply varroa treatments at the correct time.

Late summer requeening may need to be carried out, especially as varroa and miticides seem to shorten the productive life of queen bees and drones.

Then the colonies must be prepared for winter, or 'wintered down'. The correct time for wintering depends on the locality, but in areas with autumn frosts the job should be done before the frosts occur, especially if sugar feeding is needed.

When wintering down, take the following steps:
- Apply a varroa treatment if testing suggests you need to. In the acute phase three treatments may be needed: September–October, February–March and May–June.
- Make sure each colony has a good laying queen. Look especially for drone brood in worker cells or large numbers of drones present in the hive in April or May, which indicate a failing queen.
- Check that the apiary is suitably sited for winter — sheltered from the prevailing winds and not shaded, especially in the morning. Don't locate hives in gullies without adequate air drainage, where damp air and frost will lie in winter. The apiary must be in a flood-free spot that will be accessible in spring.
- Examine hives for American foulbrood before wintering them down (Fig. 7.10).

- Some beekeepers like to strip each hive down, clean the floorboard and check it for cracks that might let in wasps or mice.
- Remove any parts of the hive that will need repair during the winter.
- You should normally unite any colonies with bees occupying fewer than six frames, as these are unlikely to survive the winter without special attention. Wintering nucleus colonies with young queens is possible, but the bees must cover all the frames before clustering and have two good frames of honey. Check them for stores in early May and again in early August.
- Late-summer absconding swarms can occur from varroa-infested colonies. These are rarely worth the effort of collecting, as they are usually small and need treating for varroa and may need a lot of feeding to get through the winter.
- Check the colonies for winter stores. The amount of honey needed for winter and early spring depends on the type of winter weather in your area, and whether there are any significant nectar sources that flower through the winter. A safe guide is to allow 16–20 kg for a normal colony, which is equivalent to seven to nine full-depth or nine to 11 three-quarter-depth frames. These stores will not all be in full frames of honey; some will be in part-frames around the areas of brood on the combs. Honey left on the hive is not wasted, but is recouped several times the next season by having a stronger colony. Leave plenty of honey on as an insurance against colony starvation, and you will have to do less feeding next spring. If there is not enough honey in the hive, feed thick sugar syrup or frames of honey from another hive that has been checked and found free of American foulbrood.
- Pollen is also needed during winter, especially to enable the winter bees to produce food for the queen and any brood. In most parts of New Zealand autumn pollen shortage is not a problem, but in some areas spring pollen is scarce either because of a shortage of sources or because continuous bad weather prevents foraging.
- A colony should have the equivalent of two full frames of pollen. If it has less, transfer a pollen comb from a hive with surplus (after checking for AFB) or feed a pollen supplement or substitute.
- The arrangement of the frames in the hive is important. Hives are normally wintered in two full depth or three three-quarter depth boxes. Bees cluster most readily on empty frames that have been used for brood rearing, so when preparing a hive for winter arrange most of these frames in the centre of the bottom box, with pollen and then honey on the outside.
- Put any remaining brood combs, especially those still containing brood, into the centre of the second box. Arrange the bulk of the honey stores on either side of these brood frames. Bees naturally create this arrangement themselves and extensive frame manipulation is seldom necessary.

- Move any combs that need culling (Fig. 7.11) to the side of the hive, especially in the bottom box. They will be empty in spring and can be removed to be rendered down.
- Don't leave the queen excluder between boxes of the hive, as the cluster will move upwards during winter and may leave the queen behind. The excluder can be stored under the lid over winter and is then in a convenient place for feeding dry sugar in spring. It may need to be placed in hot water or carefully scraped to clean off any excess of wax.
- Hives may need some top ventilation. This can be provided by a screened upper entrance in the hive lid or, failing that, by sitting the inner cover on matchsticks to allow warm, moist air to escape.
- Restrict the hive entrance to exclude wasps and mice. If the floorboard has deep risers add an entrance reducer. A floorboard with risers less than 12 mm high should keep mice out, but not wasps. If wasps are a problem the floorboard can be fitted with a tunnel entrance (Fig. 7.12), which wasps will not readily enter.
- Control weed growth in the apiary.
- Check that fencing is adequate to keep out stock, or strap the hives to prevent damage if they are knocked or blown over (Fig. 7.13). Even in a fenced apiary, if you don't use straps put a heavy rock on each lid to prevent them being blown off in blustery spring weather.
- Make sure the apiary is registered and identified with your apiary code number. Leave your name and telephone number with the landowner in case the hives are damaged by wind, stock or floods.

Fig. 7.11 A damaged comb that might be moved in preparation for culling.

Fig. 7.12 Reversible tin combined entrance reducer and tunnel entrance.

Fig. 7.13 A tidy apiary in winter.

WINTER

Honey bees don't hibernate in winter, but each colony forms a cluster (Fig. 7.14). Bees in the cluster eat honey and generate heat through muscular activity to maintain the temperature of the cluster.

When the air temperature falls below 18 °C bees begin to cluster together on brood combs. As the temperature continues to fall more bees join in, until at about 14 °C the whole colony has gathered together. The cluster maintains the temperature of brood at 33–35 °C, though the temperature at the centre of a broodless cluster is much lower.

Bees regulate the temperature inside the cluster by altering its size — as the air temperature rises the cluster expands, and when the temperature falls the cluster contracts, reducing the surface from which heat is lost. During extreme cold the bees forming the outside shell of the cluster bury their heads and thoraxes inside the cluster with only their abdomens protruding. The temperature of this outside shell is always kept around 8 °C even if the air temperature inside the hive is below zero. Bees on the outside regularly change places with those inside the cluster.

The practical limit to contraction of the cluster is the absolute minimum space that the bees must occupy. There is another limit too, as the bees may contract into such a tight cluster that they lose contact with their food reserves. To keep generating heat they require food, but if they are unable to reach stores in the hive they can die of 'cold starvation'. Fortunately the prolonged low temperatures that cause cold starvation are rare in New Zealand.

It is important to understand that bees do not maintain their temperature by warming the hive around them. Bees stay warm by clustering together and, as long as they have adequate food and can cluster on empty combs, they can stay warm even when the air inside the hive is quite cold. You don't need to keep the inside of the hive warm, and in fact ventilation is very important to allow the water vapour given off by the bees to escape from the hive so that it does not condense and make the hive damp, which is detrimental to the colony.

In New Zealand, winter is the quietest time of the beekeeper's year. With adequate hive preparation in autumn no visits will be necessary during these months. In particular, avoid opening hives from late May to early August. The time for the first spring check depends on the area and on the amount of food left in the hive, but is generally around September. A check for food reserves will be needed in August for nucs and hives overwintered in single brood boxes.

Most hives that fail to survive the winter are those that run out of honey stores. Bees don't freeze to death in winter, they starve to death assuming that varroa treatments have been correctly applied.

Fig. 7.14 Bees clustering in winter.

Seasonal management plan

August

- Prepare for new season's work.
- Get queen-raising equipment if you are going to rear your own queens.
- Assemble feeding equipment and supplies of sugar.
- Check grass spraying or cutting gear.
- Assemble frames for new season and have wax or plastic foundation on hand.

September

- Apply a varroa treatment if surplus honey flow is anticipated within eight weeks.
- Check all brood frames for American foulbrood.
- Feed if necessary.
- Spray or cut vegetation around the hives.
- Stimulate drone hives and starter and finisher hives for queen rearing.
- Hives can be split late in the month or when there are plenty of adult drones present.
- Unite any weak or queenless hives with stronger queenright hives, especially if you prefer not to increase hive numbers.
- Prepare for queen-raising programme.

October

- Apply varroa treatment if surplus

honey flow is anticipated within eight weeks, or hives are showing mite damage, or there are more than 40 mites per 300 bees after a sugar shake test.
- Remove entrance guards.
- Feed if necessary.
- Check pollen stores and feed supplements if required.
- Check all brood frames for American foulbrood.
- Control swarms.
- Requeen hives with mated queens or own queen cells.
- Split hives.

November

- Remove any varroa strip treatment products applied in early September.
- Check that treatments have worked, especially if using organic treatments.
- Feed.
- Pollen check.
- American foulbrood check.
- Rear and mate queens.
- Swarm control.
- Super up hives.
- Requeen hives.

December

- Remove any varroa treatment products applied in early October.
- Feed.
- Manipulate hives.

- Introduce nucleus hives.
- Check supers for wax moth.
- Super up.
- Prepare honey house equipment.
- Harvest and extract early crops, especially if in the city.

January

- Check surplus supers for wax moth.
- Super up.
- Extract honey.

February

- Test for varroa mite levels and treat if necessary, especially if in acute phase.
- American foulbrood check.
- Remove honey before applying varroa treatments.
- Extract honey.
- Late summer queen rearing.
- Check for wasps.

March

- Test for varroa mite levels and treat if necessary.
- Extract honey.
- Requeen hives.
- Check for wasp damage.
- Sell or store honey crop.
- Store honey supers or return to hives.

April

- Remove any varroa treatment products applied in February.
- Apply varroa treatments if necessary.

- Prepare hives for wintering down:
  - feed check
  - American foulbrood check
  - scrape surplus wax from hive parts
  - check bottom boards and fit entrance reducers
  - replace rotten hive parts or tape up any holes to minimise robbing by bees or wasps
  - Control weed growth and check hives are protected from stock
  - apply mouse bait if necessary.

May

- Test for varroa mite levels and treat if necessary.
- Remove any varroa treatment products applied in March.
- Feed sugar syrup if needed.
- Winter hives down.
- Bring in honey supers stored on hives.
- Sort combs before storage.
- Freeze combs for wax moth control.

June

- Render down wax.
- Make up new equipment for coming season.

July

- Remove any varroa treatment products applied in May.
- Make up new equipment for replacement or increase of hives.

Feeding bees

The honey bee's diet is made up of nectar, honey, pollen and water. This chapter is mostly about providing sugars for bees, and unless otherwise indicated the term 'feed' means feeding sugars in some form. Feeding pollen and giving water to bee colonies is covered at the end of the chapter.

WHEN TO FEED

Spring

Spring feeding is first and foremost about ensuring the bee colony does not starve. Spring feeding maintains colony development during bad weather, when nectar supplies are insufficient, or when honey stores are running low.

Spring feeding is also done to stimulate colony build-up — to ensure that colonies have good populations of foraging bees for pollination or the beginning of the main honey flow. If a colony is stimulated too late it will miss the beginning of the honey flow or pollination period. Stimulating a colony so the population peaks well before the target date will result in extra feeding being needed to maintain the higher population, and also increases the likelihood of the colony swarming.

Starvation is probably more common in managed hives than in unmanaged ones, because beekeepers remove a large portion of the stored honey that would otherwise be left for the bees. Dividing colonies also weakens hives, and careless manipulations may cause some colonies to be robbed out.

Swarms, nucleus colonies and packages may all need feeding after installation and until they build up enough to become self-sufficient. The feeding of queen-rearing colonies is discussed in Chapter 12.

Autumn

Autumn feeding is essential for light colonies that would otherwise starve during winter or very early spring. Colonies may go into winter light in stores because of a poor honey flow, or because too much honey has been harvested. Small colonies, such as an absconding swarm due to varroa or a late split (nucleus), will also need feeding.

Autumn feeding should be done early, when there are sufficient older bees to ripen the sugar into honey stores. The weather also needs to be warm enough for the colony to be able to process the sugar, but feeding should not be done too early so that unwanted brood rearing is stimulated. Don't continue to feed colonies past mid to late May, as winter feeding will cause the bees to store syrup in brood combs that should be left empty for them to cluster on. Late feeding can also take a toll on the physiology of the bees, which need to be in peak condition to survive the winter months.

HOW MUCH FOOD DOES A COLONY NEED?

A common cause of spring starvation is inadequate preparation of hives for winter. A two-storey colony probably needs about 30 kg of honey to survive from the end of the main honey flow to the beginning of the same flow in the following year. In some parts of New Zealand there is little nectar available outside the main honey-flow period, so either this amount of honey will have to be left on the hive in autumn, or some extra fed back in spring.

You should leave enough honey on a hive in autumn to avoid the need for feeding early in spring. In most districts 16–20 kg (the equivalent of seven to nine full-depth frames) will be sufficient. Less honey can also be left on hives if the beekeeper is prepared and able to feed sugar. Under most conditions a colony with a lot of extra feed left on in autumn (say in three boxes) will use more feed than if the honey was taken off and fed back in spring. In some areas, good winter or early spring honey flows mean that less winter feed needs to be left on the hive, provided the weather is suitable for foraging in spring.

Honey stores in a full-sized colony should never be allowed to drop below the equivalent of three frames. A heavy full-depth frame contains about 2.3 kg of honey, though a good average to work to is 2 kg per frame.

It is hard to give fixed rules on how much to feed a colony at any particular time, as this varies with weather, nectar sources in the area, bee strength, the strain of bee and amount of brood. It also varies between hives — in a single apiary some hives may be starving, yet others will have a surplus of food. Many hives in city locations never need feeding at all, because of a year-round abundance of nectar sources.

But if a hive needs feeding it should be given four to eight litres of sugar syrup at a time. Many beginner beekeepers make the mistake of underfeeding hives, and don't appreciate how much sugar actually goes into a 'good feed'. As a guideline, a strong colony can take up four litres of syrup overnight, and this would last 10–14 days in spring. A frame of honey may be consumed in less than a week if there is no nectar flow on. Dry sugar is used more slowly and is less stimulating to a colony than syrup.

The timing of feeding is very important, especially in spring. 'Little but often' will keep a colony ticking over, while 'feast or famine' feeding interrupts development. If in doubt it is always better to err on the generous side. A colony given too much syrup will store it, but a colony given too little may starve and produce nothing. Carry out a feed check every three weeks.

Not all beekeepers are perfect with their timing, and you may find yourself with a starving hive. Such a colony will be completely empty of stores, and bees will have their heads in brood cells and be so weak that they fall off the combs.

You need to act urgently. If the colony is less than several good-sized handfuls of bees, it is probably not worth saving. A large colony should be given a frame feeder of warm syrup next to the cluster, or inverted containers containing about four litres of syrup should be positioned above it. A frame of feed honey could be used, provided you expose the honey by scraping away some of the cappings with a hive tool. The frame must be put next to the bees and not on the outside of the brood box. Put an entrance reducer on the hive to minimise robbing, and check the colony again in a week's time.

TYPES OF FEED
Honey
Feed honey

Some beekeepers allow for spring feeding by storing feed honey from the previous season and distributing it as needed the following spring. Feed honey provides very good bee feed, but it will spread American foulbrood widely if taken from a diseased hive. Before taking feed honey off a hive check the hive thoroughly for this disease.

Feed honey supers should ideally contain 10 Hoffman frames so the feed frames fit into a brood box easily. Take only fully capped frames to avoid drips of honey, and use good-quality frames. These frames go into the brood nest, and can be replacements for broken combs culled in the spring.

Boxes of feed honey should be stored on drip trays or division boards in a dry, cool place. Cover them to keep out rodents, ants, bees or wasps. Leave feed honey on the hives until the first frosts start, and then move it into storage to prevent wax moth development (see Chapter 13 for advice about wax moth prevention).

Some beekeepers leave a feed box of honey on a two-storey hive all year round, and find that a half-depth, cut comb or three quarter-depth super is ideal for this. In most years no extra feeding will be necessary and there is no extra cost after the first year. One disadvantage is that the feed super must be removed every time you inspect the hive. This system suits those beekeepers registered as organic producers or who wish to adopt organic principles, or where access for vehicles is difficult in wet spring conditions.

Fig. 8.1 A capped frame of honey suitable for bee feed.

Extracted honey

Extracted honey is rarely used for bee feed. It can be mixed with warm water to make syrup but its aroma readily excites robbing. American foulbrood spores can survive for many years in honey, so don't accept cheap or free offers of feed honey from acquaintances or damaged containers from shops. As a rule, do not feed any honey to bees unless you can personally determine that it is free of American foulbrood.

Waste honey

Honey that has been burnt in a wax melter or has fermented in a tin is probably better used for bee feed than thrown away. It may cause minor dysentery in the bees, but this is not likely to be a serious problem. Test a small amount first.

Sugar syrup

Either white (A1) or raw sugar can be used to make syrup for feeding bees. Don't use brown or yellow sugar or molasses, which may cause dysentery, and don't use other sugars without first checking that they are suitable. Some sugars that taste sweet to humans are not attractive to bees, some are not nutritious for them, and others (such as lactose) are even toxic. Glucose will plug up the bees' stomachs and can be used only in a mixture such as one part glucose to three parts of white sugar. Fructose can sometimes be purchased as a liquid or powder, and is suitable for feeding to bees.

For maintenance feeding, as in autumn, use only heavy syrup of two parts sugar to one part water (by weight). Mix the sugar with warm or hot water to dissolve it more easily. A simple way of making small batches of syrup is to fill a container three-quarters full with sugar, top it up with boiling water and stir at intervals until all the sugar dissolves. Wait until the syrup is lukewarm before feeding it to bees.

For stimulation feeding, for example to accelerate colony development, or of queen-rearing colonies, use lighter syrup of equal parts by weight of sugar and water.

Surplus heavy syrup will keep for many months in closed containers, although raw sugar syrup may ferment after a few weeks. On page 278 there are some useful tables for calculating the correct proportions of sugar and water to make syrup, and to determine how much honey will be stored as a result.

Dry sugar

Feeding dry sugar is convenient, especially for hobbyists. It is easier to feed dry sugar than syrup as no mixing is necessary and no feeders are required, and it does not promote robbing in apiaries. Dry sugar is a maintenance feed and does not greatly stimulate brood rearing — use it to keep an expanding colony going rather than to rescue a colony on the brink of starvation.

Raw sugar is the most suitable for dry feeding. White sugar becomes hard as it absorbs moisture and is then difficult for bees to work.

HOW TO FEED BEES

Honey

When you feed frames of honey, simply remove an empty frame from the brood nest and replace it with a full one. Put the honey just outside the pollen comb near the cluster, but on no account put a frame of honey into the cluster as this will divide it in two. Keep boxes of feed honey covered while you're working in the apiary to prevent robbing.

Syrup

To feed syrup you need a container or feeder. Whatever the design, the feeder should be leak-proof to guard against wastage or robbing by bees and wasps.

Syrup feeding is best done in the late afternoon or early evening, especially in urban areas, to minimise the possibility of robbing. Be careful not to spill syrup on the ground.

Frame feeder

This is a hollow container, the same shape as a frame but a little fatter, which is put into the brood nest in place of one or two frames (Fig. 8.2). Frame feeders are commonly made of moulded plastic, which is better than plywood or hardboard as these must have glued or sealed joints. Wooden feeders eventually develop leaks and must be continually resealed. Cheap frame feeders can also be made by folding sheet metal into a container and attaching it to a wooden top bar.

Frame feeders hold four to eight litres of syrup, depending on their width. It is important to place floats such as pieces of wood or polystyrene chips in the feeder to stop bees drowning in the syrup. You can also stuff it with bracken fern, plastic bird netting or a piece of wire gauze. Frame feeders are easy to use as the feed is always accessible to the colony, and the feeder can be filled with a watering can (with the rose removed). Many beekeepers leave frame feeders in the brood nest all year round.

Fig. 8.2 A wide feeder and eight frames in the brood nest.

Inverted container feeder

This method uses a screw-top or friction-lid container with holes in the lid, which is placed upside down where the bees can reach it. Old coffee tins and preserving jars work well. Some commercial beekeepers use 10 litre buckets with tight-fitting lids, and heat-weld or glue a section of fine stainless steel mesh over a 50 mm hole cut in the lid of the bucket. Inverted-container feeders are easy to make and are refilled with little disturbance to the colony.

The size of the holes in the lids is critical, as the syrup must not be able to run out after the container is inverted. The surface tension of the syrup holds it in the container, and bees come up to draw syrup rather like working at a teat. Frame nails or panel nails are useful for making the right-sized holes, which should be 1.5 mm in diameter. Five to 10 holes per lid are common.

The container is filled with syrup and placed upside down over the top bars above the cluster, protected inside an empty super (Fig. 8.3). The main disadvantage of this method is the need to use an extra super and a special inner cover or hive mat with a hole in it.

Don't put very warm syrup in one of these feeders as it may run out — wait until it has cooled before feeding. When the container is first inverted some syrup will come out anyway, so spill this over the bees inside the hive and not on the ground. Remember to strap the hive or put a suitable rock on the lid, as the empty super is easily blown off.

Often beekeepers using a jar feeder will underfeed their colonies. Make sure you feed the equivalent of four litres at a time or over several feeds in quick succession. Some beekeepers use feeders with just one or two holes so the uptake of syrup is very slow. This gives maintenance feeding rather than stimulation, and has an effect like feeding dry raw sugar.

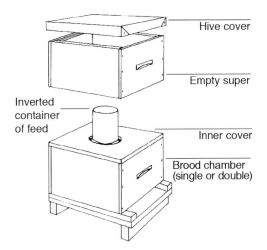

Fig. 8.3 Inverted container feeding method.

Entrance feeder

Entrance (or Boardman) feeders (Fig. 8.4) are made of an inverted container, which fits into a wooden base that sits in the hive entrance. Glass preserving jars or clean plastic food containers work well. Bees can enter the base of the feeder from the hive and take syrup from the jar, the lid of which is pierced with five to ten 1.5-mm holes in the same way as described above. Entrance feeders may be available from some bee equipment stockists.

Fig. 8.4 Entrance feeder and glass jar of syrup.

The amount of syrup remaining in an entrance feeder can be seen at a glance and the jar refilled without disturbing the colony. However, the disadvantages are that the feeder jar is small in capacity; is easily knocked out of the hive entrance by stock, and if the sun heats the jar the syrup may run out, which can initiate robbing. Bees have to leave the colony cluster to work this type of feeder, which they may not be able to do during very cold spells. Probably the only use for entrance feeders is in home apiaries.

Top feeder

This type of feeder (sometimes called a Miller feeder) is widely used by commercial beekeepers. Plastic models are available commercially. A wooden top feeder can be made from a box or half-depth super by fitting a hardboard bottom into a saw cut 8 mm up from the bottom, to provide a bee space over the tops of the frames in the box below. This makes a watertight box, provided all joints are well glued or sealed, which can hold around 10–15 litres of syrup.

Fig. 8.5 Top feeder with central chimney and plenty of flotation material.

There are two common methods for giving bees access to the feeder. Wooden feeders are often fitted with a 'chimney' — a solid block of wood with a hole bored through the middle, which is fixed over a hole in the hardboard base of the feeder (Fig. 8.5). As an alternative, bee access can be provided by making the syrup compartment about 50 mm short of one end. Bees can come up from the hive through the end entrance and feed on the syrup. With top feeders you must provide flotation material such as bracken, plywood floats or plastic bird netting, or bees will drown in the syrup.

Other methods of feeding syrup

Top feeders can be also be made by cutting suitable plastic containers in half lengthwise and placing them loosely, or nailing them, in an empty super. Flotation material must be provided, and a base of hardboard will prevent the bees sticking down the feeders to the frames beneath.

Some beekeepers try to feed bees by sprinkling syrup into empty combs from a watering can or pump nozzle fitted with a shower rose. This can waste a lot of syrup and is likely to cause robbing. If any of the empty combs are from diseased hives, the excess syrup sprinkling back into the syrup tank will be contaminated with American foulbrood spores. Frames that are later filled with the contaminated syrup will spread spores to other colonies, which may contract American foulbrood. This method is not recommended.

Various systems exist for feeding syrup using plastic bags, either stapled around a frame or filled like a bladder and punctured with holes. These may provide a cheap short-term feeder, but are unlikely to last more than a season or two, depending on the thickness of the plastic.

Never feed syrup by placing it in containers in the open. This will create severe robbing problems in the apiary and a serious public nuisance in the vicinity, and the hives that really need the syrup will not get sufficient as they are often the weakest hives.

Dry sugar

Feeding dry sugar is easy and requires no special equipment. Simply pour dry raw sugar on a layer of newspaper above the top of the brood chamber, in the bee space provided by the inner cover. This space can be enlarged if necessary by placing a queen excluder over the newspaper and then putting the inner cover on top.

Fig. 8.6 Pouring dry sugar between the outer frames, with newspaper in between the boxes.

Don't let the newspaper hang down the sides of the box, or it will draw water in and speed up decay of the woodware. Punch a couple of slits in the newspaper with a hive tool to allow the bees up, but do this before the queen excluder is put on or you might bend the wires. The dry sugar absorbs moisture given off by the cluster and is easily worked by the bees. The disadvantage with this method is that you can waste a lot of sugar if you need to inspect the colony before the bees have consumed all of it. Do not feed dry sugar over newspaper if you have a pollen trap on the hive, especially the underfloor type, as the chewed newspaper will end up in the trap.

Dry sugar can also be fed on top of a division board containing an auger hole, or even in frame or top feeders.

If you need to give a hive an extra large feed, place a half-sheet of newspaper between two brood boxes, on one side so that it covers three to five frames only. Pour the sugar down between the outermost two frames and the wall on the side of the brood box that the newspaper is on (Fig. 8.6). Shake any bees off the outermost two frames first. Using this method, you can inspect brood frames in the top brood box without losing any leftover sugar.

Don't pour sugar onto the floorboard, as this is wasteful and the floorboard is an unhygienic place for the bees to feed. It also attracts ants.

Water

Bees use water for diluting honey and sugar syrup before it is used for food, and also to cool the hive. In most parts of New Zealand bees can collect water all year round. They do this from moist soil or droplets on leaves, as well as troughs or ponds, especially if they contain water plants.

If an apiary is in a very dry place, or in an urban area where water-collecting bees might disturb neighbours, provide water in a container that has flotation material (wood shavings, polystyrene chips or legal aquatic plants) for the bees to walk on, or by letting a tap drip very slowly onto well-draining soil.

Pollen

In some areas and in some years colonies may run short of pollen, especially in spring when demand is high and bad weather restricts foraging. Beekeepers can see this quite clearly in the hive as pollen reserves dwindle or even become non-existent. One major effect of this, though hard to observe directly, is the shortened lifespan of workers caused by them using their own body protein to maintain brood rearing.

Beekeepers usually overcome this problem by moving hives to an area with good pollen sources. One possible option is to feed pollen-filled combs from other hives, but remember that this can transfer American foulbrood. Old pollen also may not be attractive or nutritious to bees.

Another method of overcoming pollen shortages is to use a pollen trap to collect pollen in areas or times of abundance, and feed it to deficient colonies. Spring is the best time to collect surplus pollen, but leaving traps on too long can denude a colony of pollen for its own needs and may hasten queen supersedure. As the new queen may not be able to leave or re-enter a hive with a pollen trap, the colony is in danger of becoming queenless. Trapped pollen can also spread American foulbrood, so colonies used for trapping must be regularly checked for this disease.

If the pollen needs to be stored before feeding, the simplest way is to freeze it in airtight containers. Fresh or recently frozen pellets can be fed by pushing them into empty combs placed inside the hive, or mixed into syrup. Feeding pollen by placing it out in the open can be wasteful, and will benefit strong colonies much more than weak ones.

If natural pollen is not readily available, a substitute can be fed, but remember that while research has improved the quality and acceptability of substitutes, there is still no perfect replacement for pollen.

Branded pollen substitute products are advertised in the *New Zealand Beekeeper* magazine. You can also make a mixture yourself. The most commonly used recipe is the 'Beltsville Bee Diet', so named as it was developed at the US Department of Agriculture laboratory in Beltsville, Maryland. The recipe is: 12.5 kg (half a bag) of fine-grain, ring-dried lactalbumin (Alacen), 25 kg of deactivated brewer's yeast and 70 kg of white sugar. Sodium caseinate (Alanate 180) or wheat protein may be substituted for the lactalbumin.

You can also add disease-free trapped pollen to the mix to increase attractiveness — even five or ten per cent helps greatly — but be sure it was harvested from hives checked carefully for American foulbrood.

Add water carefully to the dry ingredients to reach a consistency that will make firm patties — if you put in too much water you will need a lot more of the dry ingredients to get back to the right consistency. Do the mixing outside as blending the fine powdered ingredients can make a mess.

Fig. 8.7 Feeding pollen substitute over a broodnest.

Make patties weighing about 0.5 kg and press them between two sheets of waxed paper. These can be frozen until needed. You can also store the bulk mix in an airtight bucket and trowel it onto the top bars in hives as needed (Fig. 8.7). This recipe is for commercial beekeepers using bags of product, and hobby beekeepers are not likely to need this amount of substitute.

Beekeepers have to evaluate the benefits of feeding pollen substitutes in their own areas, but where pollen shortages are known, feeding supplement two or three times during the spring build-up can be very productive.

You can keep bees without doing any supplemental feeding, but for productive bee-keeping — and sometimes just to stop colonies starving to death — most beekeepers find that regulating the food supply for their colonies is an important part of colony management.

> ***Sugar syrup feeding***   To stimulate colonies during queen bee rearing, use light syrup made of equal quantities by weight of dry sugar and water. For all other feeding, use heavy syrup of two parts sugar to one part water. Mix the sugar with warm or hot water to dissolve it more easily. See page 278 for detailed recipes.
>
> ***Pollen substitute***   Combine 12.5 kg lactalbumin or sodium caseinate, 25 kg deactivated brewer's yeast, 70 kg white sugar plus some disease-free trapped pollen if available. Add water carefully to the dry ingredients until firm patties can be formed.

9 Swarms and swarm prevention

Swarming is the honey bee's instinctive method of reproduction and dispersal, so it is hard for beekeepers to control. The most common type of swarming is where one colony divides from another and the parent colony continues in existence. Another kind of swarming is absconding, where a whole colony deserts a hive. This has been observed in New Zealand only since the advent of varroa.

By swarming, honey bees in the wild are able to increase colony numbers and survive natural disasters such as disease, starvation and climatic changes. Honey bees are also able to colonise new areas by repeated swarming. In New Zealand varroa has eliminated most of the dark feral colonies that previously inhabited bush and rural areas. Swarms from managed colonies continue to repopulate some of these sites, though such colonies will survive only two or three seasons before varroa eliminates them as well.

In New Zealand most swarming occurs from late September to early December, before the main honey flow starts and when large amounts of brood are being raised, the adult population is increasing rapidly, and food supplies may be erratic. Absconding swarms due to varroa generally occur from late December to April.

THE SWARMING PROCESS

Swarming begins when the queen lays in queen cell cups, small structures shaped like acorn cups that are made by workers on the combs. Fertilised eggs laid in queen cell cups are raised by the workers as developing queens. These swarm cells are different from the types of queen cell made when the queen needs replacing, usually because of old age (supersedure cells) or when the colony suddenly becomes queenless (emergency cells).

When a queen is nearly ready to emerge from a swarm cell, preparations for swarming reach their peak. Half or even more of the worker bees stop foraging and, along with some drones, gorge themselves with honey and leave the hive with the old queen. This is called a prime swarm, and usually leaves the hive during the middle of the day. The bees pour out of the hive, fly around in a great mass for a while before settling a short distance away (Fig. 9.1). This behaviour should not be confused with orientation flights of young bees, which can occur in large numbers on a warm afternoon after a period of inclement weather.

While the swarm clusters, scout bees fly out and search for a dry cavity suitable for a nest site. On finding one, the scouts return to the clustering swarm and perform a dance to indicate its location. The cluster will usually fly off to the nest site and take up residence there within a day or two. Any swarms that stay in the open to build comb rarely survive the winter in New Zealand.

In the parent hive the first virgin queen to emerge from a swarm cell grooms herself, and then usually proceeds to tear open the sides of the other queen cells, possibly even stinging the occupants. Worker bees complete the task of destroying the cells and removing the developing queens. Should two virgins emerge at the same time they will usually fight until one queen stings the other. A queen's sting is not barbed, so this act does not kill the queen using her sting.

A new virgin queen will fly from the hive in about six to 10 days in the autumn, or two to four weeks in the spring, undertaking orientation flights at first. She may make several mating flights and mate with several drones. A few days after mating she will start to lay eggs.

Sometimes the first virgin to emerge after swarming does not tear down the remaining queen cells, but instead leaves the hive with another swarm. This is known as an after-swarm, and is usually much smaller than a prime swarm. Occasionally a hive will throw out a number of afterswarms in quick succession and end up with a mere handful of bees and a virgin queen. The afterswarms may be only the size of a tennis ball. If this happens late in the season, the parent colony and the afterswarms rarely survive the winter.

CAUSES OF SWARMING

There is no single cause of swarming, just as there is no single method of swarm prevention. Swarming is stimulated by a complex combination of factors, none of which on its own will usually result in the phenomenon. The main contributor, though, is a reduction in the pheromones or 'queen substance' workers receive from the queen, as these pheromones inhibit queen cell construction.

Worker bees receive less queen substance because the queen produces less as she ages. They also each receive less in a crowded, congested hive, as in spring, because congestion in the brood nest interferes with the transmission of pheromones.

Other factors that contribute to swarming include:
- rapid influx of nectar and pollen
- the genetic makeup of the colony — some strains of bee swarm more than others

Fig. 9.1 Swarm clustering temporarily on a branch.

- a heavy infestation of varroa mites, which can lead to absconding, especially in late summer.

SWARMING AND THE BEEKEEPER
In the days of beekeeping with fixed-comb hives such as box hives and skeps, swarming was welcomed as a way of obtaining colonies, to replace any that had died or been killed at the previous honey harvest, or to increase colony numbers. The swarming season was the highlight of the beekeeping year, as beekeepers hurriedly caught and hived swarms.

Under a modern system of beekeeping, swarming is extremely undesirable. It greatly reduces a hive's honey production or pollinating ability, often to the point of making the hive uneconomic by reducing income to below the cost of inputs (such as feeding and mite control). Absconding swarms from varroa-infested colonies spread the mite.

When a colony swarms there is a break in brood rearing while the colony does not have a mated, laying queen, which causes a break in worker emergence three weeks later, and a break in forager recruitment three weeks after that. During swarming, the queen with her particular genetic characteristics is lost, and there is a risk that the hive will become queenless if an accident kills the replacement queen (for instance during a mating flight).

The whole swarming process occupies a colony for at least a month. Preparations for queen rearing begin two to four weeks before the first swarm emerges, and developing queens are sealed before swarming takes place. About a week before swarming, the old queen's egg-laying rate drops so that she loses enough bulk to be able to fly from the hive. A few days before the first swarm leaves, workers become quieter and much of the flying from the hive is concerned with scouting for nest sites. A few hours before swarming, the colony becomes completely chaotic as final preparations are made. All these activities mean that even if a swarm is captured and reunited with the parent colony, honey production will be less than for a similar hive that has not swarmed.

SWARM PREVENTION
It is far better beekeeping to prevent swarming preparations from being made, than to try and stop a swarm from leaving the hive. However, swarm-prevention techniques must be compatible with profitable colony management. As an eminent early twentieth-century American beekeeper, C. C. Miller, wrote in his book *Fifty years among the bees*: 'If a colony disposed to swarm should be blown up with dynamite, it would probably not swarm

again, but its usefulness as a honey-gathering institution would be somewhat impaired.'

The beekeeping literature contains many descriptions of management systems designed to control swarming, most of which are very complicated. A good swarm-prevention method should reduce swarming with as little interference to the colony as possible. Some of the better swarm-prevention measures are described below.

Regular requeening

The most effective prevention method is to requeen hives on a routine basis, at least every second year. In areas with varroa, requeening will need to be undertaken every year anyway, as the mite itself and miticide treatments seem to adversely affect the productive capacity of the queens and/or drones.

Some beekeepers requeen hives only when they begin to show signs of needing it. This 'ambulance at the bottom of the cliff' approach will not prevent swarming or other forms of queen failure.

Providing room in the brood chamber

Two brood boxes should be used to avoid overcrowding. Maintain a good supply of combs in the brood nest by regularly culling those that are full of pollen (pollen-clogged) or are otherwise unsuitable for the queen to lay in.

An easy and effective way of providing extra space in the brood nest during spring is to rotate the brood boxes occasionally. During winter the cluster moves upwards in the brood chamber and when spring comes is usually in the middle of the top box. The colony expands outwards, but will not so readily expand downwards to fill the lower box. This can make the colony crowded in the top box with an empty box underneath. Swapping the two brood boxes around gives the bees more room and they will move upwards into the empty box. The boxes can be reversed again whenever the top box seems a lot more crowded than the bottom, perhaps two or three times each spring.

Providing room for honey storage

A first honey super may be needed well before the main honey flow, especially if the bees are able to gather significant amounts of nectar from early sources such as willow, barberry, thyme or buttercup. This first super should contain drawn combs if possible, as the bees cannot draw out the foundation to provide room until the weather is warmer and nectar flows are stronger.

Subsequent supering up of the hives should be done in advance of the bees' requirements. The timing of supering to avoid swarming, and ways of making space in the honey supers more readily available to the bees, are discussed in Chapter 7.

Ensuring adequate ventilation

Sometimes during a honey flow bees will cluster all over the front of a hive, even at night.

Termed 'hanging out' by beekeepers, this behaviour can be a sign that the hive is short of space for ripening nectar and storing honey, but may also mean that ventilation is inadequate.

A lot of heat is produced in a hive, and during hot weather the bees may have difficulty in fanning enough air through the hive to keep the brood-nest temperature at the optimum of 35 °C. While honey is being ripened the hive becomes very humid, creating further need for ventilation.

If bees are 'hanging out' of their hives, you can provide extra ventilation by chocking up the bottom box to make a deeper entrance. There is no danger of robbing during a strong honey flow.

Using non-swarming strains

Different strains of bee vary greatly in their swarming tendencies — something to be aware of if you breed your own queens. Never use swarm cells from a hive for requeening or establishing new colonies, as you are simply breeding from the hives with the greatest tendency to swarm. Commercial queens are selected from low-swarming stocks. Carniolan stock have a reputation as being more prolific swarmers than Italian strains, but in some seasons both strains swarm excessively.

Artificial swarming

A very common method of preventing swarming is to split a nucleus off a hive that is strong enough to swarm. Make a top nucleus following the method described in the next chapter. When you have taken a 'top' out of a hive, push the remaining brood frames in the parent colony together to reduce the risk of chilling and fill up the space with empty drawn combs.

The parent colony has been reduced in strength, which should decrease the likelihood of its swarming. When the honey flow begins, the top can be reunited with the parent hive to produce a really strong honey-gathering unit, or taken away and used for increase or replacement.

Interchanging hives

A strong hive that is liable to swarm can be swapped with a weak one to exchange field bees, helping to even out the populations of the two hives. This works best if the colonies are working a honey flow. Be sure to check the brood of each hive for American foulbrood first.

SWARM CONTROL

Despite good intentions, not all beekeepers are able to practise the best swarm-prevention methods. Occasionally remedial swarm-control measures have to be taken, but these should not be regarded as a substitute for good prevention techniques.

Cutting out swarm cells is a common control method, especially among bee-keepers with only a few hives. Only swarm cells that have been laid in by the queen need to be destroyed. Small cell cups, shaped like acorn cups, are present in the hive at any time and do not indicate that the colony is about to swarm. To be certain of finding every cell, you need to shake the bees from all the brood frames — a very time-consuming task. Plastic frames have made this inspection a little easier as there are no distorted combs in which queen cells can 'hide', as there are with wax foundation frames.

A satisfactory compromise between detecting swarm cells and excessive effort is the '10-second swarm cell check'. This relies on the fact that swarm cells are mostly built on the bottom of the frames in the top box of a double brood chamber. For the 10-second cell check (Figs 9.2 to 9.4), simply smoke a hive and crack the two brood boxes. Slide the upper box forward a little and tilt it back so that you can look at the bottoms of the frames. Smoke the bees back from the bottom bars. If no swarm cells are visible, you can be reasonably certain there are none in the rest of the hive. If you can see swarm cells on the bottoms of the frames, you will have to strip the hive down and shake the bees off every frame, as there will probably be cells elsewhere too.

Other so-called swarm control techniques include clipping a queen's wings and putting queen excluders over the hive entrance. Many of these are little more profitable than Miller's blowing up

Fig. 9.2 The ten-second swarm cell check: crack the two brood boxes . . .

Fig. 9.3 . . . slide the top box slightly forward . . .

Fig. 9.4 . . . lift the top box, smoke the frames, and examine for queen cells.

a colony with dynamite. However, adequate swarm prevention is essential for beekeeping that is both pleasurable and profitable.

The queen cells that workers make before the colony swarms are different from supersedure cells, which are made when the colony replaces the queen without swarming.

	Swarm cells	Supersedure cells
Number	5–25	Usually fewer than 6
Position on the frame	Bottom, edges or centre	Mostly on the centre
Size	Different sizes	Generally uniform size

The 10-second swarm cell check:

- smoke the hive and crack the two brood boxes
- slide the upper box forward a little and tilt it back
- smoke the bees back from the bottom bars
- look up at the underside of the top box
- if you can't see any swarm cells, all is well
- if you can, strip the hive down and shake the bees off every frame and remove all the swarm cells.

Dividing and uniting colonies 10

Beekeepers start out very conscious of how many hives they have, but as you gain experience you will find that increasing — and decreasing — that number is a normal part of management through the year. The techniques are simple and reliable, and dividing and uniting hives can be an important component of your beekeeping.

You do, though, have to think about how many hives you want, and how many you have time to manage. You may end up with more hives than you want, through catching swarms or splitting colonies to prevent swarming.

DIVIDING COLONIES

There are many reasons for dividing a colony into two, or sometimes more, units. Re-queening often involves splitting a hive. Parts of a hive are often split from the parent hive and the divided portion (called a split, nucleus, nuc, top, or top nucleus) can be used to increase numbers or held in reserve as a replacement for a failing or dead colony. Dividing hives has become more important with the advent of varroa, which has seen commercial beekeepers reporting an increase in winter losses from an average of 10 per cent to as high as 30 per cent. Dividing a hive is also a common swarm prevention measure.

Making a top nucleus

The basic method for making up a top nucleus is as follows. First check the hive for American foulbrood. Find the queen and put the box containing the queen on the bottom board. Make sure the upper box contains several frames of capped and emerging brood, and a couple of frames of stores, and that the remaining space is filled with empty combs. You should shake the bees off several more brood frames to boost numbers in the top nucleus, as flying bees will return to the bottom box. Remember not to shake the frame with the queen.

Frames of brood taken for divisions should always contain brood that is sealed, and preferably starting to emerge. The brood should not cover more than half a comb on each side of the frame, as the new colony may not have enough bees to cover a full frame of brood and prevent chilling — especially if the nucleus is left in the original apiary where there might be more loss of bees through drifting. Frames with more brood can be selected if the nucleus is to be moved to a new apiary, as such units retain all their bees.

Separate this top box, which is your new nucleus, from the parent hive by a division board, with its entrance facing to the rear. Block the division board entrance to minimise the loss of bees (as described below under 'Dealing with drift').

Put a mature protected queen cell (described in Chapter 12) or a caged queen bee into the new nucleus — either immediately, or after a couple of hours, but never leave the top queenless for more than 24 hours. If all goes well the new queen will emerge and you will have two queenright units.

When you simply split a hive in this way the bottom unit may need another box to accommodate all the flying bees, making the hive three storeys high. So it is quite common to make a top nucleus by bringing another box containing drawn combs to the apiary. Make up the top by removing from the hive the two or three frames of emerging brood (without the queen), and two of honey and pollen stores with adhering bees, and putting them in the other box with enough empty combs to bring the total up to 10. The remaining drawn combs are put into the parent hive to replace the frames removed for the top. The parent hive consists of two boxes, and the top of one.

Be very careful not to let robbing start while you are dividing hives. Weaker splits or nuclei are much less able to defend themselves than are strong hives.

Making divisions for other purposes

As well as for making top nucs, this basic method for dividing a hive can also be used to make splits that are put onto separate floorboards, to be used as replacements or for increasing hive numbers.

Fig. 10.1 Nucleus hive.

Nucleus colonies can also be made up in the same way. For a four-frame nucleus (Fig. 10.1), use one or two frames of emerging brood, one frame of stores and one empty frame, together with bees shaken off one or two additional frames. Insert a ripe queen cell or caged queen.

Making nuclei from more than one hive

Composite nuclei can also be made up from several hives if no single hive is strong enough to split. Put brood, bees and stores from several different colonies into one nucleus box and add a protected queen cell or queen. There is usually little or no fighting with this method, as the bees are totally disorientated and many of those being transferred are nurse bees, which are less aggressive. Spraying the bees with air freshener helps to reduce any fighting. If the queen bee can't be found, then a protected cell can be placed in each unit.

Dealing with drift

If you leave a nuc on top of a hive with a separate entrance to the rear, many of the flying bees will re-enter the parent hive in a process known as drifting. To minimise this loss, block the entrance of the top nuc in such a way that the bees will be kept inside for a while and re-orientate to their new entrance.

You can do this with loosely packed green grass that eventually dries enough for the bees to dislodge it. Another way is to take a double thickness of newspaper, about 100 mm x 100 mm, and fold it in a U-shape over the entrance hole in the division board. Place the division board back on the hive while supporting the paper so it is jammed between the division board rim and the super underneath. Keep the paper tight over the hole while you place the top super back on the division board. In time, the bees will chew their way through the paper. Be careful when doing this in hot weather because of the danger of suffocating the colony.

If divisions are put on separate floorboards in the same apiary, flying bees will also return to the parent hive. As with making top splits, this loss of bees (or drift) can be reduced by loosely blocking the entrances of the splits with green grass, but if livestock have access to the apiary they may eat the grass used to block the entrance. A better solution is to move the divisions to another apiary at least 3 km away so the bees do not drift back to the parent colonies.

Another technique to reduce drift is to split one hive into two to four equal nuclei and leave them in the original apiary. If room allows, you can arrange the nuclei in a circle centred on the location of the parent hive, with all the entrances facing inwards. Bees will tend to drift equally between the new nuclei, although if the mother queen is present in one of them, that unit may attract more bees.

When you can't find the queen

Dividing a hive normally starts with finding the queen, but there are alternatives if she can't be found. The easiest way is to make sure the two brood boxes both contain brood and stores, and then put a queen excluder between them. Be sure that the excluder does not have any damaged wires. After four days, only the box containing the queen will have eggs in the brood cells, as the egg stage lasts for three days.

Another way of making a top without finding the queen is to shake all the bees into the bottom box, make up the top and set it on a queen excluder above the bottom box. Bees will very soon move up through the excluder onto the brood, but the queen will remain in the bottom box. You can replace the queen excluder with a division board and introduce the queen to the top nucleus, after waiting an hour or so to allow enough bees to move up.

If you have more than a dozen or so hives to divide, by the time you have prepared all the hives with excluders you can then go back and start substituting division boards for the excluders and introducing the queens. Sufficient bees should have gone up in the meantime.

If not enough bees have come through the excluder while you are in the apiary, or you intend to come back at a later date when your queens or queen cells have arrived, you can leave the queen excluders in for a week or two. In this case, when making up the nucleus you should select brood frames with mostly eggs and unsealed brood rather than emerging brood — so that there are still bees emerging from the comb when you take the nuc away and use it.

UNITING COLONIES

Uniting is commonly done to:

- consolidate two (or more) weak colonies that on their own would not gather much honey or winter well
- join a queenless colony with a queenright one, for instance after a queen introduction has failed
- reunite splits that were made for swarm prevention or requeening
- build up strong units early for spring queen-raising
- reduce the number of hives in an apiary
- unite a swarm with a hive, or bring two swarms together.

The best technique for uniting hives is the newspaper method, which if done correctly is virtually foolproof. Perhaps the only time it shouldn't be used is if an under-floor type pollen trap is in operation, as the trap will become clogged with chewed paper.

First check that both hives are free of American foulbrood. Kill the less desirable queen if both units have queens, and put the two hives together on one bottom board separated by two sheets of newspaper (Fig. 10.2). In very hot weather use only one sheet, and make several slits in it with a hive tool.

It is best to put the queenright unit on top of the queenless one, as this seems to increase the chances of the queen surviving. If you are uniting two hives in one yard, shift the weaker hive to the position of the stronger, so that fewer flying bees are lost.

The bees from both hives will chew through the paper and unite to form a single colony. The use of newspaper means that the release is gradual instead of sudden, and the bees are confused enough not to fight. Leave the united hive undisturbed for a week, and later reorganise the frames of brood and stores into one brood nest.

There are many variations on the newspaper method. If both queens are in roughly the same condition or can't be found, it is not necessary to kill one. You can simply put the hives together with newspaper between and let the bees sort out the situation. The queen on top probably has a better chance of survival.

Another alternative is to leave both queens alive, and unite them using a queen excluder as well as a sheet of newspaper. Both queens will probably survive and the unit will function as a two-queen colony. This is best done in late November to mid-December. Provide an entrance for the top brood box, or else drones will completely block the queen excluder. The entrance can be cut into the wooden rim of the queen excluder, or the top brood box can be raised on small pieces of wood at one end or at one corner.

Fig. 10.2 Uniting hives by the newspaper method.

The two-queen colony can be left over the honey flow, and in late summer the two brood boxes can be separated to make increase, or the excluder simply removed and the brood boxes run as one unit over the winter. It is not necessary to find the queen, as it is usually the younger one that survives. Sometimes the two queens will co-exist.

Colonies can also be united directly without using newspaper, either during a nectar flow or by spraying the bees with syrup. The latter method may incite robbing. However, the newspaper method is so simple and reliable it seems sensible to use it all the time. An even simpler method is to lightly spray air freshener onto each surface being joined: the top bars of the bottom colony and bottom bars of the top colony. The smell of the air freshener seems to make the bees mingle without fighting, and no paper is necessary.

Dividing and uniting are part of many important colony management operations, and if done correctly will almost always be successful.

11 Moving hives

New beekeepers who buy established colonies often find that moving hives is their first introduction to beekeeping. For novice beekeepers it's easy for moving hives to turn into a disaster, and this first encounter with honey bees can sometimes be traumatic.

Moving hives is an integral part of management for beekeepers whose hives have to be shifted between wintering sites and honey-production sites, into orchards and crops for pollination, or to establish new apiaries and relocate existing ones. Done properly, moving hives can be a straightforward business, but you do need to be prepared for the unexpected.

All apiaries must be notified to the management agency for the American foulbrood pest management strategy if hives are kept there for more than 30 consecutive days. The details of apiary registration are set out in Chapter 20.

MOVING HIVES SHORT DISTANCES

If hives are shifted only a short distance, bees will navigate using familiar landmarks and will return to the old hive location. There they will mill about creating a nuisance, and in the evening they will cluster near the old site and eventually perish. You should try to avoid this by moving hives at least 3–5 km away from the old site (slightly less might be sufficient in city areas).

There are ways of moving hives short distances, such as across a residential section:
- Move them no more than two metres at a time. This movement can be done at any time, but it is best done at night or when it is raining to avoid confused bees annoying the neighbours.
- If you are moving groups of hives, as long as you move them together (which helps the bees find the correct hives), you can move them a bit further each time, say four metres or so. Leave them for several days between successive moves.
- You can shift hives further if you leave one hive to catch the returning bees.

The population of this hive will obviously be boosted, and the populations of the moved hives reduced. That could aid your colony management (e.g. to help prevent swarming).

- You can also shift hives at least 3–5 km away for three to four weeks, and then back to their new site. During the period away from the area, most of the field bees that remember the original location will die. You will need to leave the hive away for several months over winter, when worker bees have a longer lifespan.

LONG-DISTANCE MOVES

Moving hives can be rather unpleasant work, and without proper preparations the exercise can become very problematic.

Always check the condition of the hives before moving them, especially the floorboard. Replace any floorboards that are rotten or have runners that are not firmly attached to the base. Do this check several days before the move, as opening hives breaks the propolis seal, which otherwise holds the boxes and frames together and prevents frames from swinging about and crushing bees.

You should move hives only when the bees are not flying — either at night or in cold or wet weather. It is possible to shift bees in the morning if the hive entrances have been screened the previous evening, or if some hives are to be left on-site to catch any bees left behind.

Hives shouldn't be shifted during the day when a nectar flow is on, as fresh nectar is easily shaken out of the combs and will drown a lot of bees. If you want to shift bees during the honey flow, wait for evening or until a couple of cold or wet days stop the bees foraging.

Take care with your preparations to avoid mishaps later. Wear full protective clothing, even at night, and have the smoker going in case a hive comes apart. Bees crawl at night and in the rain, and when seeking warmth seem to find their way under veils or into bee suits that aren't fully zipped up. You should take real care to be completely 'bee tight' before you start shifting hives, especially at night. Take another beekeeper, or at least an assistant who is well suited up. If hives are more than one box high, two people will be needed to lift them.

Most problems with moving bees are caused by hives separating and allowing bees to pour out. You can prevent this by driving four nails or U-shaped fencing staples into the floorboard just inside the risers to stop the bottom box from moving sideways. These staples or nails, which the box fits snugly over, are also useful if stock rub against hives.

You should strap hives before moving them. Several types of hive strap can be purchased from bee supply stockists, and if fastened tightly they will stop the hive parts skewing apart. A cheap hive strap can be made from a length of nylon rope with a loop in one end. Tied very tightly with a truckie's dog-leg knot, this works nearly as well as a commercially made strap. Hardware shops and bee equipment stockists sell cargo straps

Fig. 11.1 Hive with top and entrance screens strapped up and ready to be moved.

with ratchet fasteners, which can secure individual hives as well as stacks of hives on a vehicle or trailer.

If you aren't going to shift hives often enough to make buying straps worthwhile, you can also fasten a hive together by nailing timber battens down three of the sides, ensuring that they are fastened to the floorboard and each super. This method is clumsy and causes damage to the hive, and needs to be done in advance of the shift as the bees become upset by the hammering. Given the availability of cheap cargo straps, we don't recommend this method.

Before starting the move, hobby beekeepers should block the bees in the hive. For a short journey, block the entrance with a piece of wire or plastic mesh cut to the width of the entrance, but deep enough that a U-shaped fold can be pushed at least 100 mm into the hive. This prevents bees from blocking the entrance and stopping ventilation. Fasten the gauze mesh to the hive with staples, and use wide tape to block up any holes in the hive or cracks between ill-fitting boxes.

For a longer journey, any journey in a closed vehicle, or on an open vehicle in hot weather, provide the bees with some top ventilation. A top screen is simply a wooden frame with side dimensions the same as a bee box (505 × 405 mm), fitted with fly-screen mesh. Take off the hive lid and inner cover and nail or strap the screen onto the top box (Fig. 11.1). A substitute top screen can be made by placing rectangles of wire gauze or nylon mesh over the top of the hive, and nailing or strapping on a queen excluder. Alternatively, drape nylon mesh over the hives and secure it to the sides of the boxes with packing tape.

Nail battens and top screens in place before actually shifting the hives, to give guard bees time to return before you close the entrance.

Once the hives have been closed up and tightly strapped, load them onto the truck or trailer, placing them with the long sides of the hives parallel with the direction of travel. Push the hives close together and tie the load on tightly. Always tie the hives to a vehicle, even if each hive is individually strapped. Single-axle trailers bounce around a lot, and many hive loads can fall apart if travelling over rough ground. Be sure to secure any other hive parts such as lids or inner covers that form part of the load. These frequently fall off trucks and trailers.

It is good practice to cover the whole load with bee-proof mesh or scrim and tie this down in case you have a breakdown and need mechanical assistance. Carry water to wet the bees if necessary, especially on hot days or in case you have a breakdown.

Always stop after five or ten minutes' travelling to check the load as ropes may loosen, especially if the hives are driven over rough ground at the beginning of the journey. Don't stop in populated areas or at petrol stations, as any escaping bees are attracted to the overhead lights.

At the new apiary, set the hives out in their permanent positions. Put the hive lids on if travelling screens have been used, and remember to pull all entrance blocks out. You can unstrap the hives and remove top screens on a later visit if you wish. Before you leave the apiary, take one last look at the hives and double-check you have removed all entrance closures.

Fig. 11.2 Truckload of beehives securely strapped down and covered with mesh for safety.

SHIFTING HIVES ON A LARGER SCALE

Commercial beekeepers usually shift hives without any entrance blocks ('open entrance') and without top screens, but use shade cloth or scrim to cover the load (Fig. 11.2). Many beekeepers use ventilated floors as part of their varroa management, and these have the added benefit of providing extra ventilation during hive movements. Both individual hives as well as pallets can be fitted with

Fig. 11.3 Two-wheeled barrow for moving hives.

ventilated floors. Beekeepers shifting hives exceptionally long distances, including for example inter-island travel, usually employ commercial carriers with curtain-sided refrigerated trucks.

Means of loading hives vary. Many hives are still lifted onto trucks by hand, especially for kiwifruit pollination where only small trucks can navigate through the orchards. Hive barrows make the job easier, though probably not quicker, and are used in conjunction with ramps or tail-gate lifters. Barrows may be single-wheeled, double-wheeled (Fig. 11.3),

Fig. 11.4 A Bobcat-type loader shifting hives on pallets.

or even motorised. Crane-type loaders are very common and come in many variations that can lift hives singly or on pallets of four. Pallets of hives are also lifted onto trucks by forklifts such as Bobcat loaders or modified tractors.

Hives from commercial apiaries are usually loaded after dark or in the early morning. The truck is parked in the apiary with the parking lights on and motor running. The parking lights provide illumination without attracting bees to the deck area of the truck, and the vibration from diesel motors in particular calms bees somewhat when the hives are on the truck deck. Most beekeepers also have extra lighting rigged on the back of their trucks, and many use red lights as honey bees can't detect the colour red.

Hives are smoked before being lifted onto the truck, where they are placed close together facing fore-and-aft. The load is tied down tightly. Exactly the reverse procedure is followed when unloading hives at the new apiary. On warm evenings, which are common during green kiwifruit pollination, the whole front of the hives can be covered with bees when it is time to unload. Some bees can be driven back inside the hives with smoke, but usually the beekeepers just put up with the bees and the extra stings they get.

Be prepared for the move!
When shifting hives, you will need to take with you:

- smoker, fuel, paper and matches
- full-cover protective bee suit with boots and gloves
- hive tools
- staple gun
- ropes or cargo straps
- hive straps, entrance screens, top screens or scrim
- spare fuel, tool kit, puncture repair kit or tyre-repair aerosol can
- mobile phone with charged battery or vehicle charger
- food and drink for the beekeepers
- water to cool the bees if they start overheating.

Queen bees  12

The most fundamental part of managing colonies is to ensure that each one has a young, laying queen. Old or otherwise poorly performing queens will make your beekeeping much more difficult and less productive, and queen failure has disastrous consequences for the colony and its productivity, reducing or often eliminating a year's honey crop and making the hive unsuitable for pollination.

THE HEART OF THE HIVE?

The queen has long been recognised as the most important single member of a honey bee colony. Early investigators of the honey bee colony referred to the 'king bee', but in 1609 Charles Butler asserted in his book *The feminine monarchie* that the king bee was in fact a female, and so of course called her the queen.

The name changed, but not the misconception that this bee ruled or controlled the colony. Though the queen is vitally important to a colony, she alone does not determine when things are done. Worker bees have a considerable effect on a colony's activities, such as controlling the queen's laying rate, replacing her at intervals, and making preparations to swarm. Workers also determine when and to what extent drones are tolerated in a hive.

The queen (Fig. 12.1) is the mother of the workers and drones in the colony, and functions as the workers' surrogate father by carrying their fathers' genetic material in her sperm sac or spermatheca. The queen is also the producer of the 'queen substance', a mixture of chemical compounds called pheromones that bind the colony together as a cohesive unit.

HANDLING QUEENS
Finding queens

You should learn to find queens quickly.

Fig. 12.1 The queen and her court.

You can carry out manipulations such as requeening or dividing hives without finding the queen directly, but this is time-consuming and less efficient.

The queen's long abdomen usually makes her stand out on the face of the comb. However, a small queen or one that is very 'flighty' and runs rapidly on the comb, such as a virgin, is much harder to see. Dark queens such as Carniolans are also harder to find than lighter-coloured Italian ones, as they blend in with their dark-coloured offspring.

When looking for a queen, move gently and be careful not to jar the hive. Use a minimum of smoke, as any excess can stampede the bees and cause the queen to start running. However, you also need to keep the bees under control and more or less off the top bars, with frequent but gentle puffs of smoke.

Fig. 12.2 Looking for one queen among a mass of workers.

To find the queen in a double brood chamber, separate the boxes and put the top box on the upturned hive lid. Cover the bottom box with a division board or damp towel to keep the rest of the bees quiet while you inspect the top box. Remove the outside comb on the side nearest to you and scan it for the queen. It is rare to find her on this frame, but stand it near the hive entrance so that if she is there and falls off she will be able to re-enter the hive and will not become lost outside the hive.

Remove the next frame in the hive using a minimum of smoke. Before looking at it, glance at the face of the next comb in the box. If the queen is there you will often see her lifting her abdomen as she walks down the comb away from the light. It will be easier to spot her if you can position yourself so the sun is shining over your shoulder and onto the combs in the hive.

Search the comb in your hands (Fig. 12.2), looking quickly along the edges in case the queen is moving around to the other side, and then looking at the middle. Next, flip the bottom of the comb away from you and up, so you are holding the comb upside down and looking at the second side. Again search the edge of the comb first before looking at the centre.

Put the comb back in the hive, hard against the side of the box nearest to you. Repeat this procedure for the remaining combs, returning each tightly against the previously inspected comb. This prevents wide gaps between the frames that soon become filled with bees, making it impossible to easily re-inspect those frames for the queen.

The queen may be hidden in clusters of bees on the face of combs or hanging from a corner. You will need to break up these clusters by blowing on them, or gently sifting through them with your fingers.

If the search of the top box is fruitless, push all the combs back against the far side and replace the frame you took out first. Cover the top box and repeat the procedure on the bottom box, lifting it up finally to check among the bees on the floorboard. If you still cannot find the queen repeat the whole process, provided the bees have not started robbing or become too aggressive.

'Sieving' the bees should be used only as a last resort, because it can provoke serious stinging and robbing. Set an empty box on the hive stand, with the queen excluder above it but offset by about 75 mm to one side away from you. Put another empty box directly on top of the excluder and shake the bees from each comb into it. Then put the cleared comb down into the bottom box through the gap left at one side by offsetting the excluder. Do this for all the combs in the box, putting each in turn down the gap and pushing it underneath the excluder. Smoke the bees heavily from above so they move down through the excluder onto the combs. Repeat the whole procedure for the next box. The queen should be left behind on the excluder.

Marking queens

Beekeepers sometimes mark queens to make them easier to find. This is particularly common in queen-raising operations, where queens must be found frequently. Marking queens is also a useful way to check supersedure rates and to record the queen's age.

The queen is usually marked with a small drop of colour on the thorax. The marking fluid must be thin enough to stick on the queen's body, but not so thin that it runs down the sides of the thorax where it can block the spiracles (breathing holes). Commonly-used markers are enamel paint used for model aeroplanes, automotive lacquer, correction pens (such as Twink), and fingernail polish.

To mark a queen, apply the paint either with a fine brush or matchstick, or a squeeze of a correction pen. You can do this as she walks across the comb, or while you gently hold her still on the comb between thumb and forefinger. If you are using a squeeze correction pen be careful not to apply too much fluid.

If you wish to mark queens to record their age, use the international colour code for this purpose. Each year is assigned a different colour on a five-year cycle:

International colour code for marking queens

Year ending with the number	Colour
1 or 6	White
2 or 7	Yellow
3 or 8	Red
4 or 9	Green
5 or 0	Blue

Specialist beekeeping supply firms sell various aids for queen marking, including holders and coloured or numbered discs with glue coating.

Queens' wings are sometimes clipped to record their age, with left wings being clipped in odd-numbered years and right wings in even-numbered years. Clipping does not prevent swarming, but is a 'swarm control' technique of the most basic kind and should not be relied on as a management practice. Registered organic honey producers are not permitted to clip the wings of queen bees.

QUEEN-RELATED PROBLEMS

Many beekeepers replace only those queens that 'look' inadequate or whose offspring are bad-tempered. A queen's appearance is not a reliable guide to her performance, although it can give some indication. A good queen has a full and tapered abdomen, does not run too rapidly on the frames, and lays a solid pattern of brood in each comb. A queen with badly frayed wings, a hairless and very shiny thorax, a short, stubby and pointed abdomen, or one that is 'flighty' on the combs, should be replaced.

Some beekeepers leave a queen in a hive until signs of failure become obvious. One of these signs is a spotty brood pattern, which may indicate a reduction in egg-laying ability due to old age, lower egg viability because of inbreeding, or brood diseases. Queens that run out of sperm, or were never mated in the first place, are drone layers and must be replaced. Older queens are more likely to swarm or be superseded. Other reasons to replace a queen include evidence of excessive disease such as chalkbrood or sacbrood, conditions like halfmoon disorder (see Chapter 13), or lack of tolerance to varroa.

If a colony becomes queenless and cannot raise a new queen, some workers may develop functional ovaries. These workers can lay eggs but don't mate, so lay drone eggs in both worker cells and drone cells. Such brood cells usually have several eggs per cell, often deposited on the side of the cell rather than at the base. Brood produced by drone layers can sometimes be confused with halfmoon disorder or parasitic mite syndrome, which is generally associated with varroa. However, drone layers produce much more drone brood in worker cells than occurs with these other conditions.

Hives containing laying workers can be united with a queenright hive if a lot of worker bees are still present. If a colony with laying workers is fairly weak, it is better to carry the hive about 10 metres from its original position and empty all the bees on the ground. The worker bees and drones will fly home but the laying workers will probably perish.

IMPORTANCE OF REGULAR REQUEENING

For productive and enjoyable beekeeping, regular requeening is essential. Queens can live for three or four years with human intervention, but after the second year their egg production decreases and the risk of swarming increases rapidly. Requeening should be done at least every two years for the reasons given below.

Annual requeening is almost essential where varroa is present. Beekeepers with varroa report higher supersedure losses and reduced lifespan of queens. This could be due to the effects of varroa and its associated viruses, the chemical treatments used to control varroa, or the effect of varroa and chemicals on the quality and quantity of the drones. Even outside varroa areas, progressive beekeepers requeen every hive every year.

Improved varroa management

The costs of treating for varroa are very significant, and are the same whether the hive is producing a good crop of honey or just making winter stores. Strong colonies are needed to outbreed varroa, especially in spring. With strong colonies, beekeepers can delay applying the first varroa treatment, or even reduce the number of treatments required.

Less swarming

Regular requeening is the single most important swarm-prevention technique. A colony headed by a queen over two years old is almost certain to attempt to swarm during spring. On average, colonies headed by two-year-old queens swarm more than colonies with one-year queens.

More honey gathering

Colonies with first-year queens usually produce more honey than those with second-year queens (even when they do not swarm). This is because young queens are generally more prolific egg layers than older queens, so their colonies have more brood during spring and thus more bees during the honey flow.

Simpler apiary management

Apiary management is much simpler if all the hives are as even as possible, since they will all need similar manipulations during a single visit. This reduces the time spent in the apiary. One of the keys to having even hives is to requeen them all at the same time.

This also assists with varroa management. If mite control chemical strips are used, similar-sized colonies will require the same number of strips to be inserted and removed. This makes accounting for strips easier, reducing the likelihood of strips being left in hives and contributing to build-up of residues and the development of chemical-resistant varroa.

Keeping the desired strain of bees

Regular requeening reduces the amount of swarming or supersedure, and so keeps the desired strain of bees in the hive.

WHEN TO REQUEEN

Mated queens cannot be produced in winter, and because requeening during the honey flow interferes with a hive's production, most re-queening is normally done in either spring or late summer. Each time has its own advantages.

The advantages of spring requeening are:

- it produces young queens for the honey flow
- queen cells are easier to produce in the spring
- taking nucs off for requeening can be part of swarm prevention
- colonies are smaller and easier to manipulate
- there are usually no honey boxes to remove
- you can increase hives or replace losses before pollination or the honey flow
- robbing is usually not as serious.

The advantages of late summer to early autumn requeening are:

- hives are stronger
- there are more drones around and the weather is better for mating
- there is less nosema (a disease caused by parasites) and supersedure with introduced queens
- taking nucs off for requeening can be part of a hive increase or replacement plan
- failing colonies are easily identified and can be split into nucs for replacement
- nucs may be able to make enough honey for winter stores without extra feeding
- bought queens are more readily available.

QUEENS AND QUEEN CELLS
Buying

Queen bees and queen cells are available from queen producers, most of whom advertise in *The New Zealand Beekeeper* magazine (details can be found at the end of this book). Place your orders well in advance as demand usually exceeds supply, especially in spring. Queen bee producers rarely deliver small orders before November because of demand from commercial beekeepers, which means that it may be best for hobbyists to requeen in the autumn or to raise their own queen bees. Many hobby clubs pool orders for their members and can purchase queens earlier, or they make arrangements with a local commercial beekeeper to produce queen cells to order.

Fig. 12.3 Top nucleus over a hive.

Queen bees can be shipped anywhere within the North and South Islands, as there are no movement restrictions for varroa control. As far as is known, offshore islands, including the Chatham Islands and Stewart Island, remain free of varroa. Beekeepers there are advised to rear their own queens rather than import them and risk introducing varroa.

Queens are delivered in plastic mailing cages, each with about seven attendant worker bees to groom and feed the queen, and with some sugar candy in one end for food. Queen cells can also be purchased from commercial producers, and will usually be couriered in small polystyrene boxes.

As many hobby beekeepers only have weekends available for bee work, and bad weather often seems to come at weekends, you should be prepared. Make up your nucs (Fig. 12.3) before the expected arrival of the queens or queen cells. Such units, if maintained above an excluder on top of queenright hives, can be prepared up to two weeks in advance. With the nucs already made, you can put queens or cells out very quickly if the weather is suitable for only a short period. Transporting and introducing queen cells are discussed later in this chapter.

Otherwise, if it is not possible to put caged queens into a hive as soon as they arrive, you can keep them inside the house in their cages for a week or 10 days. The main dangers to queen bees stored indoors are overheating, drying out, and insecticides:

- Bees tolerate cool temperatures better than heat. Put the cages in a warm part of the house, but out of direct sunlight and not near a heater or in the hot water cupboard.

- Put a small drop of water on the mesh or openings of each cage twice daily so the bees don't dry out. Use an eye dropper for this, or wet your finger and brush the water onto the openings. Do not hold the cage under a tap.
- The queens will die very quickly if an insecticide such as fly spray or house plant insect spray is used near them. Hide the spray cans and remove any automatic fly spray dispensers for as long as the bees are in the house, so that no one uses them by mistake. Don't store the queens in a house that has pest strips in it, as these give off an insecticide that is toxic to bees. Residual insecticides applied to walls and ceilings could also be harmful to caged queen bees if the house is left locked up on a warm day.

Transporting queen cages

If you need to transport caged queens to apiaries away from home, wrap the cages in a damp cloth, and preferably put them in a polystyrene container. When you get to the apiary remember to take care of any remaining queens when you are working on a hive. Queens can easily be killed if the cages are left on a hive or in a vehicle in full sunlight. Remember that queen bees can tolerate cool temperatures better than heat.

Introducing caged queens

The queen of a particular colony produces a unique quantity and mixture of chemicals or pheromones known as the queen substance. Any new queen is recognised by workers as having a different queen substance, and she will probably be rejected by the colony if introduced directly.

You can take various precautions to increase the acceptance rate of introduced queens, such as:

- allowing the worker bees to become accustomed to the new queen's pheromones before she is released among them
- introducing the queen to young bees
- confusing the hive so much that the change in queen substance isn't noticed
- introducing a queen whose pheromone output is 'in balance' with that of the old queen, so the changeover is not noticed by the worker bees.

The usual methods of introducing caged queens are based on the first and second principles, while methods based on the other two are much less common.

A caged queen should be put into a unit that has been made queenless only recently, and especially one that is composed mainly of young bees. The queen is confined to the cage for several days by a block of candy that is eaten out by the bees. During this time the bees in the hive exchange food through the mesh or openings on the side of the cage, taking pheromones from the attendant bees and distributing them around the hive.

Method one

A good rule is never to kill a queen until you have something better to put in her place.

To requeen a hive, make up a split with emerging brood using a minimum amount of smoke, checking each frame for queen cells and destroying any present. The caged queen can be introduced as soon as the nucleus is made up — it isn't necessary to leave it queenless for a time. Inserting the queen straight away helps hold the bees in the nuc. If for some reason the split must be left queenless, this should not be for more than 24 hours.

Opinions differ on the need to remove the attendant bees from a cage before introducing it. Generally this is necessary only if the queen has been in the cage for more than a week, or if any attendants are dead. If you do want to release the attendants, take precautions not to lose the queen. If you are outside, open the cage with both hands inside a big clear plastic bag. You can also open the cage inside near a closed window, so that if the queen flies out she will go to the window and you can pick her up and return her to the cage. You can also do this inside a vehicle, but be careful as queens can disappear inside ventilation slots.

Before putting the queen cage into the hive, be absolutely sure to remove the plastic tab over the candy end of the cage. It is a good idea to keep these while requeening in an apiary, so you can check at the end that the number of tabs equals the number of cages introduced.

The position of the cage in the hive is one of the most crucial factors in determining whether or not the queen will be accepted. In a cold snap the worker bees in the hive will cluster on the brood, so that is where the cage must be put, and not between or on the top bars where it may be abandoned. It is also not desirable to put the cage close to any varroa strips or other varroa chemicals that might be in the hive.

The hive bees must also be able to communicate with the attendant bees and obtain pheromones through the mesh or slots in the cage, so it is pointless to jam the cage between the combs so the slots are covered. Before introducing the caged queen, take a brood comb out of the centre of the hive and push the side of the cage horizontally into the face of the brood just under the top bar. The exit hole in the candy end should be slightly uphill, so if any attendants die after the cage is introduced they don't roll down and block it. If there is mesh on one side of the cage only, put that face downwards so that any honey running down the comb doesn't enter the cage and cover the queen. Replace the brood frame with the cage on it in the centre of the hive, and push another brood frame up against the cage to support the other side (Fig. 12.4). You may have to take a comb or the feeder out of the hive to make room for the queen cage.

In three days' time you can make a quick check to see that the queen has been released from the cage. If on this first check the queen has been released, take the cage out and gently squeeze the frames back together to prevent the bees building burr comb in the gap. Leave the cage under the hive lid or in the entrance until further checks show the hive to be disease-free.

Don't interfere with the brood nest for another 10–14 days. Disturbing the colony too

Fig. 12.4 Queen introduction cage placed correctly in the brood nest.

early can often lead to the new queen being rejected. After 10–14 days you should make a brief examination of the combs for eggs. Their presence shows that the queen has been released and is laying. This method of introduction is usually successful because the nucleus is made up mainly of young bees. If the introduction fails, the queenless split can simply be reunited with the parent hive, which still contains the old queen.

If the introduction is successful, wait for at least three weeks, and until the new queen has some sealed brood, before you requeen the parent hive. To do this, simply kill the old queen and unite the newly requeened unit on top by the newspaper method described in Chapter 10.

Method two

An alternative method of requeening is to kill the queen in a hive and insert the new caged queen as described above. The success rate of this method is lower than that of the first method, and if the new queen is not accepted there is no back-up available to keep the hive going.

If this introduction fails the colony will usually attempt to rear its own queen, but any queen produced this way will be reared under an emergency impulse. It will generally be inferior in size, egg-laying ability and pheromone production to a properly reared queen, and the colony will suffer from having been queenless for so long.

Method three

A simpler method is to lightly spray a number of brood combs with air freshener before introducing the queen cage. As with using air freshener to unite colonies, the smell seems to confuse the bees' communication system enough to allow the queen to join the colony without being killed. This direct introduction method is less commonly used than the methods above, but is highly successful.

RAISING YOUR OWN QUEEN CELLS

Rearing your own queens can be a very rewarding part of beekeeping. To be successful you will need to learn about honey bee biology, and be a close observer of colony behaviour. This section gives an introduction to several different methods of raising queen cells. If you want to delve into the subject further, we recommend you read some of the manuals listed at the end of this book.

Biological basis of queen raising

Queens and workers both develop from fertilised eggs, and are fed similar food for the first three days of larval life. During this period the larvae are capable of developing into either female caste: queen or worker. Three days after hatching the caste of the larvae is determined by the diet they receive. Queen larvae continue to be fed copious quantities of royal jelly, while worker larvae have honey and pollen added to their food.

The basis of queen raising is to take from worker cells female larvae that are less than three days old, and encourage the worker bees to feed them copious quantities of royal jelly. With this diet the larvae will develop into queens.

Although any worker larva less than three days old can be made to develop into a queen, the best queens come from those that are selected when less than 24 hours old. This is important because queens reared from young larvae are bigger and have more egg-producing tubes (ovarioles), and therefore can lay more eggs.

Queens can be reared whenever there are adequate nectar and pollen supplies, enough fine weather for mating, and an abundance of sexually mature drones. Don't start rearing queens unless you can see drones in the hives, otherwise there won't be mature drones when the virgins emerge and fly to mate. Drone supplies are usually adequate from September to March, although this varies with season and location.

Varroa predation on feral and managed colonies, plus the increasing use of plastic frames, has meant there are fewer drones available for mating than in the past. Some of the reported queen failures and reduced productivity from queens in areas with varroa may be due to a lack of drones, especially in the spring. Progressive commercial beekeepers who have switched to plastic frames are now inserting one or more frames of drone comb in each brood nest to help ensure an adequate supply of drones.

Natural queen-rearing impulses

In nature, worker bees rear queens in one of three circumstances:
- total queenlessness or emergency conditions
- supersedure of an inadequate queen
- prior to the colony swarming.

The best artificially reared cells are raised in an initial queenless state, followed by a simulated supersedure condition. Other methods are also used but some of these give poor results.

Unsatisfactory ways of raising queens

Some beekeepers make up nuclei or splits to increase hive numbers or replace losses, but don't introduce a ripe queen cell or mated queen. The small nucleus is left to raise its own queen under an emergency impulse.

Worker bees select young worker larvae, enlarge their cells, and feed them royal jelly. In this situation the feeding is less than adequate and, surprising as it may seem, the bees are not very good at selecting the optimum age of larvae. At least one-third of the larvae selected may be three or four days old, and are on the borderline of being unable to develop into a queen at all, let alone a good one. Of course it is the virgin from the oldest larva (and thus a poor queen) that emerges first, tears down the other cells, and becomes the colony's queen. There are many absolute failures with this method, and few 'successes' result in good queens. It is not without reason that this method has been dubbed 'pauper's splits'.

It is also bad beekeeping practice to use swarm cells from a hive to make up new colonies or to requeen others. Although the cells may be satisfactory, they are all being taken from hives with swarming tendencies, so this is a good way to breed a swarming instinct into your bees.

Queen raising without grafting

There are several ways of raising queen cells that do not require grafting, which is the delicate art of transferring larvae from worker cells into queen cups. Two methods using whole combs will be outlined here. These are simple ways of raising a few cells, and they are especially good for those people whose eyesight or steadiness of hand is not adequate for grafting.

One drawback of whole-comb methods is that they involve mutilating a drawn-out comb for every batch of cells raised. The combs must be drawn from wax foundation, so plastic frames or wax foundation sheets are not suitable. Some methods also rely on the bees to select the larvae to be raised as queens.

The best non-grafting method involves the use of queen-rearing kits, which are sold under brand names such as Jenter and Nicot. They provide a way of obtaining and transferring larvae of the correct age, using a cage to confine the queen and plastic cell cups to move the larvae. You can have a high success rate raising quality queens using a kit — the main disadvantages are the high initial cost of the equipment, and getting them to work properly. When they do work well, you can raise hundreds of queen cells at a time, usually far more than are needed.

Miller method

This requires a breeder colony from which you wish to raise offspring, and a cell-building colony that is made queenless when the cells are in it. Both colonies must be fed sugar syrup, and possibly pollen, for a week or two before cell-raising. You can use the breeder colony as your cell builder, in which case you only need one colony, but you will need a nuc box or another brood box to contain the queen while you start and finish the cells in the cell builder (which must be queenless at that time). More details are given under the grafting method.

Take an empty but nearly drawn-out brood comb made from wax foundation and cut a zigzag or straight line across it part way up the comb. Discard the bottom half of the comb, and put the remaining portion into the breeder colony. The breeder queen will lay eggs in this comb. After four to six days remove the comb from the hive.

Trim the cut line of the comb back to larvae of the right age, using a scalpel or very sharp knife. It may pay to warm the knife in a flame or hot water to give a neat cut without distorting the cells too much. Place the comb in the centre of a queenless cell-building colony. The cell builder should be a strong, queenless hive that is bursting with bees, especially young ones. It is important to check that there are no queen cells in the cell-building hive when you remove the queen. The hive should have very little unsealed brood left in it, so that the nurse bees will concentrate on feeding the queen cells. Provide some extra feed.

After 10 days take out the frame, which should have a number of ripe queen cells built on it (Fig. 12.5). Cut these carefully from the comb and use for requeening, replacement or increase.

Fig. 12.5 Queen cells produced with the Miller method.

Alley method

This method also uses a breeder colony as the source of larvae and a cell-building colony. Place an empty drawn-out comb in the centre of a breeder colony. The breeder queen will soon start to lay in this comb, and after four to six days it will contain eggs and larvae of the correct age. Take it from the hive and cut strips of comb that contain the appropriate larvae (Fig. 12.6). Remove the cells from one side of the comb, cutting down to the midrib of the foundation, and mount the strips of comb onto wooden cell bars using molten beeswax as glue. Pour a quantity of wax onto the bars before you fix the strips of foundation. This makes it easier to cut the strips off when the cells are mature.

Fig. 12.6 Cutting strips of comb for raising queens by the Alley method.

When each strip is fastened to the bar, squash down the cells leaving every fourth one (Fig. 12.7). This will leave the finished queen cells separate from the others so that they can be easily removed. Put wooden bars with prepared strips of

Fig. 12.7 Squashing cells in a strip of comb for raising queens by the Alley method.

comb into their special frame, and place the frame in a cell-building colony. Use the same kind of strong, queenless, well-fed unit as used in the Miller method.

After the cell frame has been in the cell-builder for 10 days, remove it and harvest the cells by cutting the strip carefully off the bars with a sharp knife and then separating the cells. Use the thick wax base when holding the cells with your fingers, rather than the cells themselves.

Grafting methods

Worker larvae were transferred into queen cells by the Swiss naturalist François Huber in the late 1700s, but the technique wasn't practised commercially until it was developed by the American G.M. Doolittle in the 1880s. Almost all current grafting techniques are based on the Doolittle system.

Grafting involves transferring worker larvae less than 36 hours old from a well-fed breeder colony to artificial queen cells, which are put into a cell-building colony for the next 10 days. The most common technique is to place the queen cells containing the larvae into a queenless 'starter' colony for 24 hours, and then into a queenright 'finishing colony' for the next nine days, although it is common to use one hive as a combination starter/finishing colony by making it queenless first and then making it queenright again.

You can make artificial queen cells at home from beeswax, or buy plastic cell cups from a bee equipment stockist. Both are equally effective, although the plastic cups are used more by commercial beekeepers because they are much easier to handle than wax cups. Cells are mounted on wooden bars that fit into a standard frame, with a maximum of three bars of wax cells or two bars of plastic cells per full-depth frame (as the plastic cells are longer). Wax cells are fixed to the bars with wax, and plastic cells either with wax or by fitting short lengths of wooden dowel to the bars that the cups can easily fit over. A little wax also helps the cups to stay on the dowel.

If you require only 20–30 queen cells at once it is convenient to use a single colony for raising them. Larger numbers need separate colonies. For the single-colony method you will need a strong colony that occupies two brood boxes and has at least six frames of brood. In early spring you may have to unite two colonies to get a single strong colony. This cell-rearing colony can also be the breeder hive.

One week before grafting, arrange the hive so that the queen and most of the unsealed brood are in the top box, and two or three frames of sealed brood, several frames of honey and pollen, and a syrup feeder are in the bottom box. Put a queen excluder between the two brood boxes, and feed the colony with a litre of 1:1 (by weight) sugar syrup. Provide pollen either in the syrup (a cupful per 10 litres of syrup) or by mashing down a pollen comb with a hive tool.

Continue to feed the hive at two- to three-day intervals until grafting, giving only one or two litres of syrup at a time as you are simulating good nectar-flow conditions rather than trying to increase the stores in the hive. If you overfeed, the bees will build burr comb all around the queen cells.

About mid-morning on the day of grafting, remove the top box containing the queen and place it on a new floorboard, or a division board with a rear entrance, just behind the colony. At this visit you should also select and mark, with a drawing pin or felt pen, a frame in the breeder hive that contains larvae between 12 and 36 hours old.

Put the queenless unit on the old floorboard. Bees flying from the box containing the queen will return to the old entrance position, boosting the strength of the queenless unit. Place the frame(s) with empty cell bars between the brood frames in the centre of the queenless box, so the bees can clean and warm the cell cups. It may be necessary to remove one of the outside frames to make room for this. The feeder should stay with this half of the hive, which is to be used as a queenless cell-starting colony. Feed one litre of light syrup and some pollen. Check that there are no queen cells in this unit. Do not feed your breeder hive at this time or syrup will drip all over you when you are holding the frame with young larvae during grafting.

By early afternoon the bees will be in a good state for receiving the grafted queen cells. They should be 'roaring queenless' as they try to detect the queen's pheromone, and hanging in big clusters on the empty cells. Any colony that does not form clusters over the cells, and does not have bees boiling over the top bars when you lift the lid (Fig. 12.8), is not strong enough to produce the best cells.

Fig. 12.8 Strong queenless hive ready to take bars of grafted queen cells.

Fig. 12.9 Grafting larvae from worker cells into artificial queen cell cups.

In preparation for grafting, take the marked brood comb from the breeder hive and brush it clear of bees using a soft-bristled bee brush (available from bee equipment stockists) or a large feather. Grafting can be done outdoors on a fine and warm day, but in windy or cool conditions use the shelter of a vehicle or building (Fig. 12.9).

Graft in good light, but never in direct sunlight as this may kill the larvae. You might find it helpful to use a head-torch, and remember to use reading glasses if you need them.

Use either a retractable grafting tool (known as a Chinese grafting tool) or a very fine artist's brush, size 00 or 000. To remove a larva from a worker cell, put the tool or brush down the side of the cell so the tip just curls under the larva. A larva of the right age lies in a 'C' shape, and it is best to lift the larva with the tool or brush at the back of the C. Lift the tool or brush straight out of the cell, bringing with it the larva and a small amount of royal jelly.

Carefully place the larva into a queen cell cup, and either roll the brush slightly or retract the tip of the tool to pull it out from underneath the larva. It is very important not to turn the larva over in the royal jelly, as this will block its breathing holes and suffocate it. Discard any larva that has rolled, brushed against the cell wall, or otherwise been damaged, as it will almost certainly not develop.

Fig. 12.10 Placing grafted cell cups into a queenless starter hive.

Fig. 12.11 Bees clustering on queen cells.

Some beekeepers 'prime' the cell cup by placing a small amount of royal jelly in it before grafting. Priming is not necessary in spring, although it may be beneficial in summer or autumn. If you do prime the cell, be sure to use pure royal jelly from a natural queen cell (swarm or supersedure cell), and not a mixture of royal jelly and honey or water.

If you are fairly slow at grafting it may be best to prime the cells, and return the first bar to the hive when grafted while you are grafting the second. To do this you will need to have a cell frame with removable bars. If you are experienced at grafting and sitting in the right conditions, you can graft a whole frame at once.

When the whole frame has been grafted, return it to the hive (Fig. 12.10). The space for the frame will be full of clustering bees. Try to disturb them as little as possible, because these bees will go onto the cells and feed them. Give another feed of one or two litres of syrup and gently close up the hive. Return the brood frame to the breeder hive.

About 24 hours later, examine the frame to see how many cells have been accepted. An experienced queen-raiser will find 80–95 per cent of the cells being worked by the bees (Fig. 12.11), with copious royal jelly around the larvae and fresh wax being added to build the cell walls. First attempts often bring disappointing results, with perhaps less than 50 per cent acceptance. Continued practice is the only way to improve your success rate. If you are using plastic cell cups, move the accepted cells into the centre of each bar, away from the edges of the frame where they may be deserted if the weather turns cold.

At this time reassemble the hive — return the queenright box to the original floorboard, put the excluder on top of this box and then the box containing the queen cells on top of the excluder. Feed the colony again. The hive has now been set up as a queenright cell-finishing colony.

On the ninth, or preferably the tenth, day after grafting, the cells must be removed and used in hives. The cell-raising hive can then be used as an ordinary honey hive with the queen excluder removed from between the brood boxes, or it can be set up again to repeat the cycle and raise more cells. In the latter case, first destroy any extra rogue queen cells the bees may have raised on the brood combs in the top box.

There are many variations of this basic method. Some involve special tin slides or division boards that simplify the technique and reduce the disturbance to the colony.

Some beekeepers use a 'swarm box' as a queenless starter. This essentially is an extra deep five-frame nuc box. The extra deep sides are usually ventilated with flyscreen mesh, as swarm boxes have their entrances blocked when starting queen cells. Swarm boxes need to be kept in a cool place during this period, and some beekeepers use a temperature-controlled room. Swarm boxes contain one frame of honey and pollen, two frames of sealed brood and a feeder, plus space for the cell bars. The queen cells remain in the swarm box for 24 hours and are then transferred to finisher hives.

USING QUEEN CELLS
Timing

The immature queens in cells are most robust on the tenth day after grafting, and this is the best time for putting them out into hives. It is important that this is not delayed by even half a day. The queens may start emerging late on the tenth day after grafting if the larvae used for grafting were rather old, or if the weather was warm during the cells' development. If the cells are still in the finisher colony, the first virgin to emerge will tear down the other cells.

If it is not possible to put cells out on the tenth day, then it is best to do it on the ninth day, or even earlier in an emergency. Remember, though, that the immature queens are very delicate on the seventh and eighth days after grafting and must be treated very gently. Any cells that are dropped should be discarded as they will not emerge, or if they do the queens will often have curly wing tips and will not be able to fly and mate.

Transporting cells

Ripe queen cells must be protected from temperature changes, physical shock and drying out while they are being transported to another apiary. Cell carriers should be transported inside a vehicle and never on a trailer. Cell bars must never be shaken to remove bees from the queen cells, but the bees brushed off using a large feather or soft bee brush.

Perhaps the best environment for the cells is provided by a bee colony. Make up a nucleus using two frames of brood, the frame of queen cells, and one or two frames of honey from the finisher unit. Make sure it has plenty of ventilation, especially in warm weather — a screened top instead of a lid is ideal. The disadvantage with a nucleus carrier is that you need to brush the bees off each cell before using it. Many bees are lost each time the nucleus box is opened, and if visiting several apiaries, not enough bees may be left to keep the cells warm.

When you have finished with the nucleus, return it to the hive it was taken from, or leave it in an apiary with the last queen cell in it for hive increase.

Another suitable cell transporter is a polystyrene container such as a chilly bin, with a hot water bottle or a plastic bottle filled with warm water in the bottom wrapped in a damp towel. It is important that the water used is warm rather than hot, as it is very easy to overheat the cells. To protect the cells from shock, lay them between layers of soft polyester such as filling for pillows and duvets.

Be sure not to leave the cell transporter in direct sunlight or it may overheat. Temperature regulation using this method may not be as good as with a nucleus carrier, but there is no risk of bees escaping inside a vehicle and the cells are ready to use once in the apiary. If the travel time to an apiary is not great, then a water bottle warmer may not be necessary, provided the container and the polyester linings are pre-warmed first by placing them in a hot-water cupboard overnight.

Introducing cells

The rules for introducing queen cells are similar to those for introducing caged, mated queens. Cells can be introduced to a colony that has just been de-queened, but once again the best advice is 'never kill a queen until you have something better to put in her place'.

We suggest you always protect the queen cells by covering them with aluminium foil or masking tape, or inserting them into a small length of hose pipe. Ensure that the tip of the cell is clear of the covering so the queen bee can emerge. Queen bees produce a pheromone that inhibits worker bees or other queen bees from tearing down the tip of the queen cell, so cells are usually attacked through the side walls. If the sides are protected, the virgin queen will at least be able to emerge, and so will have a better chance of surviving. A variation on this method is to make the protective tube twice as long as the cell cup plus queen cell. Some beekeepers believe that this extra length provides a space the queen can retreat into if necessary after emerging.

To use queen cells, make up a nucleus from the parent hive, checking for queen cells on the several frames of mixed-age brood used, and add a few frames of honey and pollen.

Place a protected queen cell in the middle of a brood comb, just under the top bar where it will be covered if the colony clusters. Make sure that the tip of the cell is not covered, as it is from here that the virgin will make her exit. Be very gentle pushing the cells into the brood frame, especially if you have produced the cells using the Miller or Alley methods as the base of the cells will be very fragile. Plastic cell cups are much stronger so are easier to place between frames, which is the main reason beekeepers use them.

Fig. 12.12 Using polystyrene baby nucs for mating queens.

Refrain from checking on the nucleus (Fig. 12.12) for two weeks, as any disturbance during this time will increase the chances of the new queen being rejected.

A virgin will emerge from the tip of a cell leaving a neat round hole, sometimes with the cap still hanging onto one side. Occasionally the cell cap is pushed back over the hole, and it looks at first glance as if the queen has not emerged. If a cell is torn down by another queen or worker bees, there will be a jagged hole in the side of the cell. If the queen has not emerged, or if after three weeks there is no sign of eggs in the hive, first check that no emergency cells have been built, then put in another ripe queen cell. You may need to add another frame or two of emerging brood to boost the hive population.

Wait until the nucleus containing the new queen has some sealed brood before killing the old queen and uniting the new unit on top.

Queen cells and varroa treatments

Some beekeepers blame poor queen performance on the mite control strips that are often present in hives being requeened. It is probably prudent not to have strips in hives, or use formic acid or thymol treatments, after placing queen cells in colonies. If you choose to leave the strips in, then at least place the queen cell as far from the strips as possible (although still keep it inside the brood cluster). Similarly, varroa treatments should be removed when cell-raising.

Learning how to requeen a hive is, as we have said, a fundamental part of hive management. It is worth paying careful attention to the techniques outlined, as purchased queen bees are expensive and the consequences of failure even more so. Rearing your own queen bees can be a very satisfying part of beekeeping, as well as giving you more control over your requeening programme.

13 Parasites, diseases and pests

No matter how experienced you are, all the beekeeping management skills in the world will not do you any good if you fail to prevent and control pests and diseases.

In this chapter we outline diseases affecting brood or developing bees, diseases that affect adult bees, and parasitic mites. We also cover the most common beekeeping pests and other causes of bee deaths.

The more scientists study bee parasites and diseases, the more it seems that honey bee colonies can be affected by a complex interaction between parasites, viruses and pesticides. Colony loss, particularly on a large scale, often cannot be explained by looking at one agent only, and is probably caused by a particularly virulent combination of parasites and pathogens.

This chapter looks at diseases, parasites and pests from a beekeeping management perspective. More information about their science, and about important bee diseases and pests not present in New Zealand, can be found in the books and other information sources listed at the end of the book.

MITE PARASITES
Varroa mite (*Varroa destructor*)

The most significant change in New Zealand beekeeping in recent times was the discovery of the varroa mite in April 2000. For the previous 50 years, this parasite had been a major concern in countries with developed beekeeping industries, and had spread to most parts of the world through both deliberate and accidental movements of infested bees.

After initially being confined to the North Island, varroa was found in Nelson in the South Island in 2006, and after a period of containment in that area it eventually spread further. All movement restrictions on bees and bee products were removed in September 2008, so it is only a matter of time before varroa spreads throughout the South Island.

Resistance to one of the principal varroa control products, Apistan, has already been detected in New Zealand. It is not known how widespread this resistance is, but varroa has developed resistance to such compounds worldwide, so New Zealand is unlikely to be any exception. The development of resistance will make the task of controlling varroa in New Zealand that much harder.

Dealing with varroa warrants a whole book in itself, and all New Zealand beekeepers should read the excellent manual *Control of varroa — a guide for New Zealand beekeepers* by Mark Goodwin and Michelle Taylor, which is available from the National Beekeepers' Association. It covers in depth varroa biology, development, sampling and control, whereas this chapter can cover only the basics of varroa.

Varroa (scientific name *Varroa destructor*, formerly *Varroa jacobsoni*) is an external parasitic mite of honey bees that evolved in association with the Asian honey bee *Apis cerana*. It jumped species and began infesting the European honey bee *Apis mellifera* in the early 1960s.

Varroa destructor is aptly named, since the mite and its associated viruses will kill untreated honey bee colonies. Varroa has killed tens of thousands of colonies since its arrival in New Zealand, including nearly all feral colonies. Over 2500 beekeepers have given up beekeeping since 2000, and it is only 10 years later that new beekeeper registrations are exceeding the number that exit the industry each year. Most of the turnover in beekeeper numbers is in hobby beekeepers owning five hives or fewer.

Varroa is such a destructive parasite that you cannot keep honey bees in New Zealand without applying some form of chemical treatment at least two or three times per year. The treatments also need to be given at the correct time and rate. The choice of chemicals and the time of application will depend on factors such as personal philosophies about using chemicals, how long varroa has been in your area, the location of your hives, the predicted start and end of the honey flow and the amount of time you have available to invest in varroa control.

What do varroa look like?

Varroa are visible to the naked eye and look like sesame seeds. They are crab-shaped and measure 1.1 mm long by 1.6 mm across. Mature female adult mites are the ones you are most likely to see, and they are dark

Fig. 13.1 The female varroa mite.

reddish-brown to black in colour (Fig. 13.1). The males are smaller and rarely seen as they are born, mate and die in the brood cells.

Mites are usually seen walking over brood combs, or attached to drone or worker pupae that become exposed on the top bars between brood boxes when the boxes are split apart. You can also see them if you lift the pupae out of their cells.

However, do not rely on a visual diagnosis of bees or brood combs for the presence of this mite. Varroa are very difficult to see on adult bees, and it is generally accepted that if you can see varroa on the combs, the colony has a dangerously high level of infestation.

How long do varroa live?
Varroa need brood to reproduce, but adult females can live and feed on the 'blood' of worker bees in broodless hives for months over winter. We have found varroa still alive on a sticky board (Fig. 13.2) four days after it was removed from a hive, and on dead and decaying brood after three weeks.

Fig. 13.2 Stickyboard showing hundreds of varroa mites.

Fig. 13.3 Varroa on a pupa in its cell.

Life cycle
Adult female mites enter brood cells before they are capped and submerge themselves in the royal jelly at the bottom of the cell. They breathe through a special modified appendage that acts like a snorkel.

After the brood cells are capped, the mites emerge from the royal jelly and begin feeding on the 'blood' (haemolymph) of the worker or drone larva (by now at the pre-pupal stage) (Fig. 13.3). The mites lay six to eight eggs but the first one is always a male. This male mates with the new females inside the cells, but only mature and fertile female mites emerge with the bee and survive outside the cell.

Mites prefer drone brood — this enhances reproduction as drone development time is longer, and more mites mature sufficiently to survive outside the cell before the drone emerges. The reproductive rate in drone brood can be 2.6 new female mites per mother mite, twice the rate in worker brood.

Most females can produce more than one generation under New Zealand conditions where we have a long brood-rearing period. As brood is usually present, even in winter, varroa can reproduce all year round.

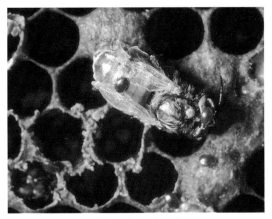
Fig. 13.4 Worker bee infested with varroa.

What effect do varroa have on bees?

Varroa feeding on bees can cause physical damage to wings and other body parts, but it is bee viruses that cause most of the harm, since varroa provides a pathway for viruses to get into the haemolymph or 'blood' of the bees. Some viruses kill bees quickly, while others may interfere with protein synthesis and take days or weeks to show any effect on a colony. There is no known cure for viruses; however, controlling varroa will usually also reduce the effect of the viruses.

How quickly does varroa kill colonies?

Once varroa is in your area, untreated colonies can die out within 12–18 months. Colony death happens suddenly, and very strong hives with maybe two or three supers of bees in late January can be empty of bees by the middle or end of February.

When a colony collapses because of varroa infestation, almost all of the bees disappear and presumably die. Often the queen stays behind with a cupful of workers, or sometimes by herself. The honey remains on the hive, with robber bees often reluctant to take it. There are no piles of dead bees out the front of the hive as you see with pesticide poisoning or starvation.

New Zealand scientists have recorded up to 122 mites entering hives per day on adult bees. This invasion is usually much worse in autumn when healthy colonies are robbing those that are dying of varroa, and carrying mites back to their hives. Invasion is worse in the first three to five years the mite is in an area, known as the acute phase of varroa. Treatments, each lasting six to eight weeks, may be required over three periods during this time: spring, mid to late summer and mid to late autumn.

Once most or all of the feral and untreated managed colonies in an area have died, varroa settles into what is called the chronic phase. Varroa populations can then be more easily controlled by beekeepers as there is less invasion — most beekeepers find they need to treat only twice a year if using miticides applied as strips, in spring and autumn. More treatments are usually required if organic products are used.

Parasitic mite syndrome (PMS)

The brood condition of colonies with heavy infestations of varroa is described as parasitic mite syndrome, or PMS. The true cause of PMS is not known. The term describes a range of brood symptoms associated with varroa, but these may not all appear at the same time. It affects larvae from C-shaped ones in the bottom of the cell to older larvae or pre-pupae lying lengthwise in the cell. PMS can occur at any time, but is more common in the late summer and autumn when mite levels are generally highest.

Symptoms described as PMS can include:

- Patchy capped brood with sunken and perforated cappings. The perforations are jagged and often off-centre as with AFB, sacbrood and chalkbrood — and the disease European foulbrood (EFB), which has not been found in New Zealand.
- Affected larvae can be seen anywhere in the brood combs.
- Larvae may be twisted up the cell wall as in half-moon syndrome (HMS) or EFB (Fig. 13.5).
- Larvae may be coiled in a half-moon shape around the cell wall and often near the lip of the cell (Fig. 13.6), as with EFB and HMS.
- Larvae and pre-pupae can be light brown or yellow in colour, as with AFB or EFB.
- Unlike with AFB, the body tissues can be lifted out of the cell and do not rope out.
- The symptoms can disappear if the colony is treated with an effective miticide.

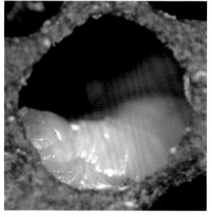

Fig. 13.5 PMS-affected larva twisted in the cell.

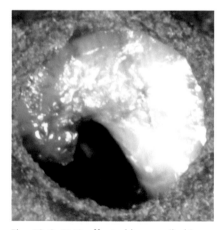

Fig. 13.6 PMS-affected larva coiled in a half-moon shape around the cell wall near the lip of the cell.

It is very important not to misdiagnose American foulbrood (AFB) as PMS, since combs with brood killed by AFB will contain large numbers of disease spores that can set off an infection in other hives if the combs are re-used or visited by robber bees. If in doubt get a second opinion from an approved beekeeper or an apiculture officer, or arrange for a larval sample to be tested at an approved laboratory. Alternatively, treat for varroa for two weeks with a proprietary miticide that has a quick knockdown, such as Apistan or Bayvarol, and then re-examine the brood frames for AFB. Use new Apistan or Bayvarol strips for this.

Sampling hives for varroa

There is a range of good sampling methods described in the book *Control of varroa* mentioned above. One of the easiest is the sugar-shake method. It is bee-friendly, and can be done on one visit to the apiary.

Fit a 500 ml jar with a plastic or wire mesh lid and fill it one-third full with about 300 bees. Push a dessertspoon of dry icing sugar through the mesh onto the bees, and roll the jar for a few seconds to coat the bees in sugar. Vigorously shake the jar over a light-coloured tray or hive lid (Fig. 13.7). Keep shaking until no more varroa appear. Most mites present will fall through the mesh with the sugar and can be seen easily.

Fig. 13.7 Using the sugar-shake method.

The sensitivity can be increased by repeating the test. Apply a second spoonful of sugar and roll and shake as described above. The bees can be returned to the colony when you have finished with them. Do not reuse the icing sugar as it will not work well a second time. It will also not work in the rain, and sometimes too if there is a very heavy honey flow, since the icing sugar gets sticky.

If you see 40 or more mites then you need to treat the hive as soon as possible. You may also need to treat if you cannot get back to your hives for another month or so, as mite numbers can explode from both natural development and invasion, particularly in the autumn.

Treating hives for varroa

This is a large topic and for full information you should study the *Control of varroa* book. You should only use approved chemicals to control varroa. Each comes with specific instructions, but none of the products should be used when there are honey supers on the hives. Chemicals used to control varroa are often described as organic (meaning the active ingredient

Fig. 13.8 Applying Apilife VAR treatment.

is found in nature, even though the chemical itself may be manufactured) or inorganic/synthetic (meaning the chemical or molecule is not found in nature).

Chemicals approved for varroa control fall into three categories:
- proprietary products with synthetic active ingredients, such as Apistan, Bayvarol and Apivar
- generic chemicals such as formic acid, oxalic acid, thymol and food-grade mineral oil
- proprietary products containing the generic chemicals, which have been prepared and delivered in a form suitable for varroa control, such as Apilife VAR (Fig. 13.8), Apiguard and Thymovar.

The proprietary products have been registered for use in New Zealand and have approved labels that specify how they are to be used. If you use the generic chemicals not as proprietary products but by buying the raw ingredients, you must follow the code of practice or guidelines established for these chemicals. These are set out in *Control of varroa*. Both formic and oxalic acid, in particular, are very hazardous materials and misuse can harm the operator as well as the bees and hive equipment.

For all products, at least two visits are required: one to apply, and another to remove the strip or container. Extra visits are needed for most of the organic products, as they may not be slow-release and may require several applications to give an effective dose without damaging the bees or brood or leaving residues.

Using organic chemicals Things to keep in mind when using organic chemicals to control varroa include the following:

- Despite the label 'organic', some acids are very hazardous to bees, beehive parts and the beekeeper if not used correctly.
- Currently no organic chemical has been found to be as effective as synthetic chemicals.
- Organic chemicals are very variable in their effectiveness, for reasons not well understood.
- Sampling to determine the effectiveness of organic chemicals is essential — otherwise you should be prepared to accept a higher proportion of colony losses.
- Honey supers must be removed and organic chemicals applied as early as possible in late January through to February. This is to allow time for monitoring to identify where the treatments have not been very effective, and for more or different treatments to be applied.
- Manufacturers' label requirements must be adhered to for best control and operator safety.
- Follow the codes of practice if you are using the pure chemical.
- Organic chemicals should generally be used in conjunction with synthetic chemicals, although some beekeepers report success using proprietary thymol-based products on an almost continuous basis when honey supers are not on the hives.
- If you are using organic chemicals you will have to make more visits to the hives to apply the chemicals. This will have an economic consideration, especially for commercial operators.
- Organic chemicals can leave residues in honey, wax and propolis, though these may reduce over time.
- Resistance of varroa to organic chemicals has so far not been detected, but in theory resistance could develop.

Using synthetic chemicals The three products registered as strip formulations in New Zealand give the longest and most consistent control of varroa (Fig. 13.9). Things you should consider when using these products are:

- Resistance of varroa to Apistan has already been reported in New Zealand. This means that resistance to Bayvarol may also occur, as both active ingredients are in the same chemical family.
- Beekeepers will now need to be extra observant when removing strips to see if varroa control has been effective.
- Take care to keep these products out of the sun, as the active ingredients can be destroyed very easily.
- None of them are registered for use in hives producing bee products under organic certification.
- Residues of fluvalinate have been found in beeswax and propolis.
- Follow the label recommendation, and do not under-treat especially now that resistance to Apistan has been reported.

Fig. 13.9 Applying mite control strips.

Table 13.1   Varroa control products registered for use in New Zealand

Name	Active ingredient	Application type	Visits required	Weeks of treatment	% varroa killed[1]
Apistan®	Synthetic pyrethroid (fluvalinate)	Plastic strip	2	8	95–100
Bayvarol®	Synthetic pyrethroid (flumethrin)	Plastic strip	2	8	95–100
Apivar®	Amitraz	Plastic strip	2	8	95–100
Apilife VAR®	Essential oils (thymol, eucalyptol, camphor and menthol)	Vermiculite wafer	3	4	65–97
Apiguard®	Essential oil (thymol)	Gel	3	6	68–98
Thymovar®	Essential oil (thymol)	Wafer	3	3–4	85–100
Generic thymol	Essential oil (crystals)	Soaked cords or crystals	3–4	3–4	66–98

Name	Active ingredient	Application type	Visits required	Weeks of treatment	% varroa killed[1]
Food-grade mineral oil	Oil	Fogging or emulsion-soaked cords	Weekly to monthly	unknown	unknown
Generic formic acid; Mitegone®	Organic acid (liquid)	Liquid in slow-release pads	3	3	25–95
Generic oxalic acid	Organic acid (crystals)	In sugar syrup	1	1 only; more can damage brood	30–40

[1] If the kill is less than 95% then the effectiveness of the treatments is either low or variable and follow-up monitoring should be carried out after treatment. If mites are still present some weeks after applying new Apistan® or Bayvarol® strips, resistant varroa may be present and a test for resistance should be carried out.

Table 13.1 is adapted from *Control of varroa — a guide for New Zealand beekeepers* by Mark Goodwin and Michelle Taylor.

Colonies must be treated in both spring and autumn. As colonies should not be treated while there is a honey flow, most beekeepers count backwards from the normal time their honey flow starts so they are sure to leave enough time for the spring treatment. It is helpful to treat colonies as late as possible in spring, so the autumn treatment can be done after honey has been harvested. This gives better control through the winter.

The timing of autumn treatment is best decided by monitoring mite levels and making decisions on the circumstances of each season. Organic products are best applied in February, although this can be delayed if varroa levels are still relatively low.

Chemical resistance Chemical resistance occurs when a noticeable proportion of the varroa population can tolerate the chemical being used. This is a particular problem with Apistan and Bayvarol, and as the active ingredients in these products are in the same chemical family, mites resistant to one will also be resistant to the other.

Resistance to fluvalinate (Apistan) has already been detected in New Zealand, and is likely to become increasingly common. Some beekeeper practices encourage the development of resistance, such as not alternating chemical groups, leaving strips in hives well past the approved removal time, and using fewer strips than recommended by the manufacturer.

Testing for resistance If varroa are seen after the recommended treatment period for a given chemical, then resistance is possible. There are other possible causes. For instance, the chemical used might have been ineffective (e.g. because the product was left in the sun, or for some reason was a substandard batch), or it may be slow-acting (e.g. organic products). Other possibilities are that an incorrect dose was used, or very heavy reinvasion by mites took place.

If you suspect resistance to Apistan you should test the bees for resistance. The US Department of Agriculture test (www.ars.usda.gov) has been adapted for New Zealand conditions. The test involves confining bees in a jar with a small piece of Apistan strip, and counting the number of dead mites after 24 hours. The bees are washed in methylated spirits to remove any mites not killed by the Apistan, and the results calculated as a percentage.

The testing procedure is described more fully in *Control of varroa*, but in summary is:
- Take a 500 g jar, as used for testing for varroa using the sugar-shake method. Staple a 9 mm x 12 mm piece of Apistan strip to the centre of a piece of card and place the card in the jar with the Apistan strip facing inwards. Add a sugar cube.
- Shake bees from one or two brood combs into an upturned hive lid, bucket or large plastic bag. Scoop up 1/4 cup of bees (about 150 bees) and place them in each jar, being careful not to include the queen bee, or to damage the bees. Screw the mesh lid on the jar.
- Place the jar in a warm, dark room for 24 hours. Do not cover the lids, as this will cause the bees to overheat and possibly suffocate. The bees need to cluster on the Apistan strip. Remember not to use pesticides during this time, either from handheld aerosol cans or automatic dispensers.
- After 24 hours, hold the jar about 100 mm above a piece of white paper and turn it so the mesh lid is facing downwards. Hit the base of the jar with the palm of your hand three times, and count the number of mites that fall on the paper. This is the 'initial kill' used later in the calculation.
- Take the jars outside and knock the bees to the bottom of the jar, or kill the bees first by placing the jar in a freezer. Remove the index card with the attached strip and store it in an opaque or dark container kept out of the sun. You can use it a couple of times more. Half-fill the jar with methylated spirits (meths), then remove the mesh lid and replace it with the original solid lid. Shake the jar vigorously for five minutes.
- Remove the solid lid and replace it with the mesh lid. Pour the meths into a bucket through a funnel lined with a paper towel. Refill the jar half way with meths and swirl the bees around and pour the meths through the paper towel again.
- Take the paper towel out of the funnel and count the number of mites recovered. This is the 'final kill' figure in the calculation.
- If the total number of mites recovered (initial kill plus final kill) is fewer than 10, there were too few mites on the bees and the test should be repeated.
- To calculate the percentage of mites killed by Apistan, divide the number of mites that initially fell on the white paper before the bees were killed, by the total number of mites (which is the white paper mite count plus the paper towel mite count, after washing). Multiply this by 100 to get a percentage.

- Percentage killed by Apistan $= \dfrac{\text{initial kill} \times 100}{\text{initial} + \text{final kill}}$

- If the Apistan killed less than 50% of the mites, the mites may be resistant. If this is the case, talk to an AsureQuality apicultural officer to discuss options.

Living with varroa

Experience gained in New Zealand since the advent of varroa means we can draw the following simple conclusions about varroa management.

- Colonies that are not treated will usually die.
- Chemical treatments must be alternated to slow the development of resistance by varroa.
- Resistance to Apistan has now been reported in New Zealand, so beekeepers should monitor the effectiveness of both Apistan and Bayvarol.
- Bayvarol and Apistan are in the same chemical group so should not be alternated with each other. Either one of these has to be alternated with an organic product or Apivar.
- Bayvarol and Apistan contain synthetic pyrethroids, and along with Apivar have been used with success during the acute phase when effective and prolonged control is required.
- Organic miticides can be effective but are more variable than the products containing synthetic chemicals. The reasons for this variability are not well understood, but constant monitoring for effectiveness is recommended when using these products.
- Treating with organics should be done earlier in the autumn, to give time to identify any hives not responding to the treatment and to treat them again if necessary.
- In general chemicals with more reliable activity should be used in the autumn, and those with more variable activity should be used in the spring.
- Using any product contrary to label directions will promote varroa resistance.
- Managing hives with varroa is time-consuming. Commercial beekeepers report they can now manage only 250 to 700 hives per labour unit, whereas before varroa they might have managed 500 to 1000 hives per labour unit.
- The best operators practise annual requeening, as queens seem to fail quicker than before the advent of varroa. This could be due to the effects of miticides on developing queens and drones, or the lack of drones now that many beekeepers are using plastic foundation and nearly all the feral colonies have been killed off by varroa.
- Chemical residues in beeswax and propolis have been reported, and are becoming a significant issue for the industry. Carefully following the instructions that come with the miticides will reduce residues.

BROOD DISEASES

The significant diseases of honey bee brood found in New Zealand are American foulbrood (AFB), sacbrood and chalkbrood. AFB is a much more serious problem for beekeepers than sacbrood or chalkbrood, which are discussed here mainly for identification purposes to avoid confusion with AFB. Halfmoon syndrome or disorder (HMS) will be discussed briefly as it can also be confused with AFB, as can parasitic mite syndrome (PMS), which has been discussed earlier in the chapter.

No bee disease found in New Zealand is known to harm humans, or to make honey or other bee products unfit for eating.

Inspecting colonies for disease

Inspecting brood for disease should be a normal part of your colony inspections. To inspect a brood box, push the frames away from you, then pull the nearest frame towards you to make a gap. Work your way through the frames containing brood.

Before inspecting a frame, shake the bees off it inside the hive. Then look at cappings and inside cells to detect brood symptoms. Use the sharp corner of your hive tool to pick the caps off sealed cells so you can look closely at the cell contents. Do not be afraid of doing this — the bees soon repair the damage, and the brood in the cells will survive. Remember to wear the appropriate glasses if you need them for reading.

To be able to recognise diseased brood you must be familiar with the appearance of healthy brood in all stages. Healthy larvae have a glistening, pearly-white appearance and distinctive body segmentation. They lie coiled in a C-shape at the bottom of the cell (Fig. 13.10) for four days after hatching from the egg, and then lie stretched out along the length of the cell (Fig. 13. 11). After another day or so the

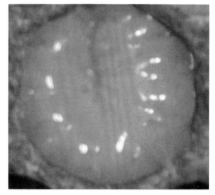

Fig. 13.10 A healthy larva at the C-stage.

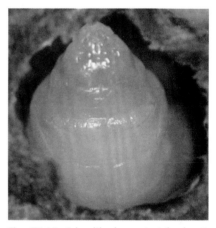

Fig. 13.11 A healthy larva stretched out along a cell.

Fig. 13.12 Domed capping of healthy brood.

cell is sealed over with a slightly convex (or domed) wax capping (Fig. 13.12). Development of the bee into a pre-pupa and pupa continues within the cell (Figs 13.13–13.14).

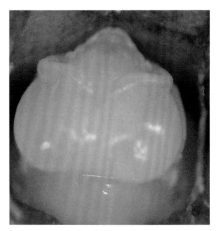

Fig. 13.13 A healthy pupa.

Fig. 13.14 A healthy pupa after body parts have begun to take on colour.

American foulbrood (AFB)

This serious brood disease is caused by a bacterium that produces very resistant, long-lived spores. A colony with AFB will usually be killed by the disease. However, this may take many months and in the meantime the bees, stores and hive parts can infect other colonies. Before a colony dies, it can have an infection of AFB that either cannot be seen or is easily missed, and the infection can be spread by interchanging equipment or bees. Other colonies can also contract the disease by robbing. Despite this, a beekeeper who takes the right precautions can reduce the incidence of AFB to extremely low levels or eliminate it. Beekeepers must be knowledgeable about the disease and constantly inspect for it.

The AFB inspection and management techniques outlined in this chapter are detailed in the manual *Elimination of American foulbrood disease without the use of drugs*. All beekeepers should obtain a copy of this book, which is available from the National Beekeepers' Association.

Causative agent

American foulbrood is caused by a spore-forming bacterium, *Paenibacillus larvae larvae*, or *P. larvae*. Some beekeepers still refer to the disease as BL or *Bacillus larvae*, the old name for the bacterium, and in this chapter we often call *P. larvae* spores 'AFB spores', which is also common beekeeper shorthand.

The disease cycle starts when *P. larvae* spores are eaten by a honey bee larva with its food. The spores germinate in the larva's gut, and the resulting active (or vegetative) stages of the bacterium multiply rapidly, invading the body through the gut wall. The developing bee will die at the larval or pupal stage. Young larvae are much more susceptible to infection than older larvae.

Spores are again formed in the decomposing body of the dead larva or pupa, and are available to infect another host, thus repeating the cycle. Approximately 2500 million spores are contained within a dead larva or pupa.

The spores of *P. larvae* are resistant to boiling water, drying out and chemicals, and have been known to stay viable for over 50 years. This obviously poses some challenges for controlling the disease.

Symptoms: colony

Any colony is at risk of contracting AFB. Strong colonies may be at greater risk of contracting AFB than weak ones, because of their superior ability to rob out weak, diseased colonies. Swarms can also carry AFB.

The foul smell that gives the disease its common name is not a reliable way of detecting the disease. A characteristic smell does develop with the disease but only in advanced cases, by which time visual symptoms are very obvious. In any case, people differ in their ability to detect this smell. A colony with a lot of brood that has died from another cause, such as chilling, starvation or sacbrood, will also have an odour of decay.

When you are examining a colony for symptoms of AFB, take care to cover scraps of wax and honey to prevent robbing. Lift the cappings back from any suspect cells using the sharp corner of a hive tool or a matchstick.

Characteristic signs of AFB are:

Patchy brood pattern   In a healthy colony headed by a young queen, the pattern of capped brood in the comb is usually fairly regular. Where AFB is present there will, except in light infections, be scattered patches of capped brood surrounded by cells that are either empty or have unsealed brood in them, as the disease has prevented some bees from developing and emerging (Fig. 13.15). In advanced cases, the brood is in a very patchy or 'peppered' pattern.

Fig. 13.15 The beginning of a patchy brood pattern in an AFB-affected hive.

There are other causes of patchy brood pattern such as a failing or inbred queen, and other brood diseases or conditions such as sacbrood, chalkbrood, half-moon syndrome or parasitic mite syndrome, but it is a characteristic of AFB so should always be treated with suspicion.

Perforated cappings Nurse bees often chew the cappings from cells containing diseased brood. This results in a fairly small hole with jagged edges somewhere in the cappings (Fig. 13.16). These holes should

Fig. 13.16 Early AFB infection, with sunken, slightly dark cells with puncture marks and one visible larva in the early stages of decomposition.

not be confused with cappings that are in the process of being completed over healthy brood, which have a smooth round hole that is always in the centre.

If you notice patchy brood, or sunken or perforated cappings, make a close examination of the brood.

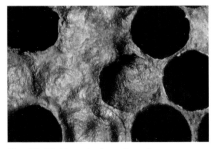

Fig. 13.17 Very dark and sunken cappings of an advanced AFB infection.

Fig. 13.18 A larva killed by AFB beginning to slump down in the cell.

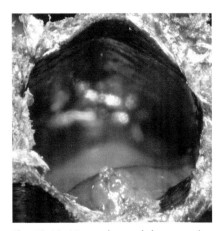

Fig. 13.19 More advanced decomposition, with the larva losing segmentation and changing colour.

Sunken cappings Cappings that the bees have made over healthy brood are usually slightly domed. Cappings over a cell containing a larva or pupa that has died from AFB are often sunken (Fig. 13.17), and may become oily-looking and darker in colour if the disease has been present for a long time.

Symptoms: brood

Although honey bee larvae become infected with the AFB bacterium at a very early stage, they do not die until either just before or just after the cell is sealed — at the late larval (pre-pupal) or at the early pupal stage. Brood that has died from AFB will have some or all of the following features:

Decay and colour change Infected larvae or pupae slump down to the lower side of the cell (Fig. 13.18), losing signs of body segmentation and changing colour throughout the body from off-white through yellow to light brown (Fig. 13.19). The corpses decay further until, about a month after death, the dead larvae form almost black scales that stick tightly to the lower wall of the cells. These scales are rarely removed by the bees, and so are useful in making a diagnosis.

Ropiness test Diseased brood initially has a watery consistency, no different from that of healthy brood. As the larvae or pupae decay further and become brown in colour, the remains become sticky and able to form a string or rope when drawn out.

To test for ropy brood, push a dry matchstick or twig with a rough surface into a cell, stir the contents slightly, and *slowly* pull the stick out (Fig. 13.20). Diseased brood draws out as a smooth, even-coloured chocolate or coffee-brown thread about 10–30 mm long. The thread is elastic and will snap back when it breaks. This symptom is characteristic of diseased

Fig. 13.20 The ropiness test for AFB.

brood at this stage. Leave the match in the comb or burn it in the smoker, as it will be contaminated with the AFB bacterium.

If there is a heavy infection of chalkbrood fungus as well as AFB, the AFB-infected larva may rope out less than described. Sacbrood may also rope out 10–15 mm but the contents will generally be mottled in colour, lumpy in consistency and uneven. The thread is not elastic. Brood with PMS can rope a little but there will be other obvious signs of varroa mites and varroa damage.

Mouthparts If the brood dies at the pupal stage, its mouthparts will point upwards to the top of the cell and may even be attached to the opposite wall of the cell (Figs 13.21, 13.22). This can be a useful diagnostic feature, since exposed mouthparts are not present in any other disease or condition, so you should take care when removing caps from cells not to destroy these mouthparts.

This characteristic sign of an AFB infection is not very common however, and is usually found when the disease has reached an advanced stage — the lack of prominent mouthparts does not mean that AFB is not present.

Smell When the disease is advanced, a characteristic foul, fishy smell *may* be noticeable.

Getting a second opinion
If you are uncertain of your diagnosis of American foulbrood, obtain advice from a beekeeper approved under the PMS, or contact the PMS management agency or an AsureQuality apiculture officer. Arrangements can be made to send a sample of suspect larvae to a diagnostic laboratory for confirmation, and the testing will be paid for by the AFB pest management strategy.

Since the advent of varroa and its associated parasitic mite syndrome, many beekeepers are mis-diagnosing AFB, often with severe consequences.

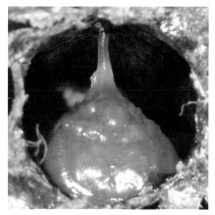

Fig. 13.21 The tongue of a developing bee killed by AFB at the pupal stage.

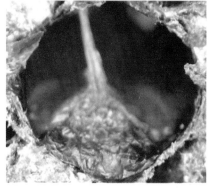

Fig. 13.22 Any pupal tongues are visible after the remains have dried to a scale.

Untreated varroa will kill colonies, including those also infected with AFB. This in turn may lead to bees robbing infected honey, or equipment being reused instead of being burnt. Correctly diagnosing AFB is important to contain the spread of this serious disease.

Destroying hives infected with AFB

The law requires that colonies and hives with AFB must be burnt, unless other actions are specifically permitted. A positive diagnosis that must result in the destruction of colonies and hives is either: identification by the owner, an approved beekeeper or authorised person; or positive diagnosis of the AFB bacterium from a brood sample.

A positive lab test for AFB from honey or adult bee samples is not deemed to be positive for AFB disease, because bees can carry AFB spores on their bodies but not in sufficient quantities to start an infection in the colony. However, if you receive a positive test from bees or honey you should mark the hive(s) in the apiary providing the sample, and increase the number of frames inspected for AFB and the frequency of inspection. It is also good practice to quarantine such hives and not interchange bees or honey or hive parts for at least 12–18 months.

Approved beekeepers will have a disease elimination conformity agreement (DECA) with the PMS management agency (see Chapter 20 for further explanation). The DECA will have provisions describing how the beekeeper is to deal with any hives infected with AFB. This can range from total destruction by fire of all hive parts, to destruction of bees, frames and hive parts that cannot be heat-sterilised. Approval may be granted in the DECA to salvage hive woodware in good condition by sterilising for 10 minutes in hot paraffin wax heated to 160 °C.

If local fire restrictions prevent the burning of hives on site, then permission is usually granted to mark the hives with 'AFB', kill the colonies and move the hives to secure, bee-proof storage before burning at a later date. This can also be approved in the DECA.

The safest time to destroy diseased colonies is late evening or during wet weather, when there is less likelihood of bees flying. Check whether there are any local fire restrictions before burning infected colonies.

Killing the bees An effective method is to block the hive entrance carefully and ensure that the hive is bee-tight by taping up any holes in hive boxes. Then raise the hive lid and pour about half a litre of petrol over the bees and combs. Quickly close the hive and leave it undisturbed until the petrol fumes have killed the bees — about 10–15 minutes. Do not light the petrol.

Burning diseased bees and combs Dig a hole about one metre or more in diameter (depending on the amount of material to be burnt) and about half a metre deep, close to the infected hives.

Part of the hole should be made deeper than the rest to allow any liquid honey that escapes the fire to be buried where it is unlikely to be disturbed. This 'sump' in the bottom of the fire hole also aids the burning, as it reduces the amount of honey in the fire pit and helps prevent it suffocating the flames.

Carry the hive parts to the hole, using the upturned lid as a tray to prevent dead bees falling to the ground. For a small hive, stack the hive parts in a loose pile so that plenty of air will be let in to the fire. Be careful when lighting the fire because petrol will be trapped in any empty combs. It is safest to stand upwind from the fire hole, and either throw in a ball of burning

Fig. 13.23 Destruction of hives infected with American foulbrood.

paper or pour a petrol 'fuse' along the ground for several metres and light that.

If the hive is large, it is best to start by arranging a few frames in a lose pile or in teepee fashion, then set them alight with a length of rolled-up newspaper. Once this is alight (Fig. 13.23) you can add more parts to the fire as it burns.

Beeswax and dry hive boxes burn very vigorously, and flames from a three or four-storey hive can reach three metres or more in height. Take the usual precautions when lighting a fire outdoors. You should obtain a fire permit if necessary and have fire extinguishers and damp sacks handy.

The hive will take several hours to burn completely. Keep lifting up the burning material in the fire pit with a long-handled shovel or fork to expose fresh timber that may be buried beneath unburned honey and wax. Plastic frames create lot of smoke, so burning this material is best done at night if possible.

Once all the hive material has been consumed, the pit should be filled in. Dig over the ground at the hive site to bury any remaining dead bees and replace the grass sod over the fire pit if possible.

Diseased hive equipment should not under any circumstances be dumped in a landfill or transfer station.

Cleaning equipment in the field

Any hive tool used in association with infected colonies should be scorched in the firebox of the smoker (Fig. 13.24) or burnt

Fig. 13.24 Sterilising a hive tool in the firebox of a smoker after inspecting an AFB-infected hive.

with a gas flame. The wooden parts of the smoker bellows should be scraped with a hive tool, and scrubbed or washed in soapy water to remove any adhering honey and bacterial spores. Gloves and any clothing or boots that may have honey or wax on them should be similarly washed. Washing this way will not kill the AFB spores, but you will at least dilute them and wash them away.

Gloves and hive tools can also be soaked for one hour in 0.5% hypochlorite solution to kill any bacterial spores. This solution can be made from taking a product like Janola (bleach) that contains 3% hypochlorite, and diluting it 5:1. Wash hands, overalls and gloves later in hot, soapy water to remove any hypochlorite. Hypochlorite is a surface steriliser, so it will not be effective on lumps of wax or propolis where the surface might be scraped off later to expose infected material. Hypochlorite will damage leather gloves over time. Follow label directions when using bleach solutions, and be careful not to breathe the fumes or splash the product in your eyes.

Sterilising hive appliances

Non-plastic hive parts in good condition may be salvaged and sterilised in hot paraffin wax if this is approved in your DECA. However, as explained above, frames, bees, honey and old appliances must be burnt.

Salvaged gear can be sterilised by immersion in paraffin wax at a temperature of 160 °C for 10 minutes. Information on the sterilisation of hive parts and appliances is available from the PMS management agency or AsureQuality apicultural officers, and is also well covered in the AFB elimination manual.

Natural development of AFB in a honey bee colony

To understand how to prevent and control the spread of AFB in your beekeeping operation, you first need to understand how AFB develops and spreads naturally.

Colonies can contain small numbers of AFB spores but not have any diseased larvae or pupae. These might be called contaminated colonies, rather than diseased ones. Colonies become diseased or infected when there are enough spores in the colony to start killing brood. Honey bee colonies have some defence against initial infections of AFB, and through hygienic behaviour are able to remove small numbers of diseased brood. In this situation the beekeeper might not know that the colony is infected, with what is called an inapparent or subclinical infection.

The problem is that contaminated colonies or ones with inapparent infections can later develop full clinical cases of AFB, by which time AFB spores are likely to have been spread by the beekeeper. This is why AFB prevention and control focuses on reducing, or at least tracking, the movement of hive equipment.

Spread of AFB

There are many ways in which AFB can be spread between colonies. The most significant means are those that shift large numbers of AFB spores, and these are common bee-keeping practices:

- moving extracted honey supers between hives, often the season after the honey is harvested
- transferring brood between colonies.

Other important causes of AFB spread are:

- beekeepers transferring empty frames or frames of food stores from diseased hives to disease-free hives
- beekeepers feeding pollen that has been collected from diseased hives
- beekeepers drying out cappings over hives, if honey has been harvested from diseased colonies
- bees robbing out diseased colonies that are weak or dead, whether colonies in hives or feral colonies
- beekeepers using hive equipment that at some time has been contaminated with AFB spores
- well-meaning people putting out honey and water syrup for birds, or spraying it onto fruit trees to attract bees, if the honey is contaminated with AFB spores.

There are other ways that AFB can spread. The significance of all of these is not well understood, but it is still worth taking preventive action to minimise risks from them:

- Drifting of bees between hives can spread AFB spores, and initiate infections. If colonies are lightly infected the rate of spread through drift is probably not high, but is still worth preventing. Drifting of bees from more heavily infected colonies is more likely to spread AFB. Methods of preventing drift are discussed in Chapter 2.
- Swarms can carry enough AFB spores to initiate infection when the swarm establishes a colony. If you hive a swarm you should treat it as potentially contaminated until proven otherwise.

Beekeeping management to reduce and eliminate AFB

The incidence of AFB in a beekeeper's hives is almost entirely within the control of the beekeeper. Through proper management it is possible to greatly reduce, or even eliminate, AFB from your hives. Worrying about other beekeepers or feral hives is missing the point, and diverting attention from the real causes of AFB.

The two most important things you, as a beekeeper, can do to combat AFB are:

- thoroughly inspect brood before removing anything from a hive
- reduce the exchange of equipment between hives as much as possible.

Inspections The more frequently you inspect brood for AFB, the more likely you are to detect the disease before it can spread to other colonies. You should inspect brood for AFB at least monthly during the beekeeping season (say September through to May), and you should inspect for AFB before taking anything off the hive, particularly honey.

Brood inspections should be thorough. For instance, if there are 12 frames of brood in a hive and you inspect only three, you have a 75 per cent chance of missing any AFB present. Inspecting more frames, including from both brood boxes, is time-consuming, but it greatly increases the likelihood of you finding AFB before it has a chance to spread to other colonies.

Remember to shake the majority of bees off a frame before inspecting it for brood disease. If you employ staff, be sure to train them in brood inspections and monitor their thoroughness.

Trace-backs It is often difficult to inspect brood nests thoroughly in late summer or autumn, as hives are heavy with honey and robbing can start very quickly. However, any honey boxes that are removed from diseased hives and used the following season can easily infect another hive.

Marking honey boxes is not a substitute for inspecting brood when harvesting honey, but can be a helpful back-up. It involves marking each hive and honey box when the harvesting is done. Beekeepers usually use a letter code for the apiary and a number for the hive within that apiary (e.g. A1, A2 ...; B1, B2, etc.). A felt pen is satisfactory, as the numbers need to stay legible for only a few months. If AFB is found during a thorough brood inspection in autumn after the honey is harvested, or even the following spring, then that hive and all the boxes removed from it can be destroyed. Some beekeepers use bar codes, hand-held scanners and computer programs to record the movement and current location of all hives and associated equipment.

Quarantine Quarantine — isolation to reduce the chances of disease spreading — is a standard practice for preventing the spread of diseases between animals (including humans). It also works to prevent the spread of bee diseases such as AFB, and can be applied at either the hive or apiary level.

Hive quarantine is labour-intensive and requires careful management, but is a very effective way of overcoming a high incidence of AFB. To implement this, each hive is permanently identified with a number or code (animal ear tags nailed to the floorboard are effective). Once a hive has been identified, no equipment is swapped to another hive. Smaller items such as feeders and queen excluders are stored on the hive. Equipment that is removed, such as honey supers, is identified with the unique hive number and used again only on that hive.

When extracting honey you will need to take care to ensure the extracted frames go back into the same box. You can either mark each frame with the hive number, or use different colours of spray paint across the top bars and on the box to ensure that frames are not mixed up and put into the wrong boxes. Each box will, of course, carry the unique hive number.

If AFB is found in any hive, any stored equipment with that hive's number is destroyed or sterilised using an approved method. This may seem harsh, but is the only way to prevent that equipment initiating an infection when it is placed on another hive.

Hive quarantine is more practicable for beekeepers with smaller hive holdings, but is a very effective tool for dealing with a high incidence of AFB.

As its name suggests, apiary quarantine involves managing each apiary separately and ensuring that no equipment is exchanged between apiaries. Apiary quarantine is used by many beekeepers in New Zealand, regardless of AFB incidence. It is practical to use it on a large scale to prevent AFB spread as well as to control outbreaks.

To operate an apiary quarantine system, beekeepers mark all the hives and equipment in an apiary with a unique apiary code number or letter. In the same way as with a hive quarantine, once equipment has been marked with the apiary code, it either stays in that apiary or is returned there.

Other management techniques to reduce disease spread AFB is spread mostly by beekeepers. Apart from the quarantine techniques described above, there are other precautions you should take to break the transmission cycle of the disease:

- do not acquire hives or equipment unless the seller is registered with the PMS management agency as required by law
- treat swarms and feral hives as potential sources of disease
- do not use feed honey of unknown origin — if you have not checked the hives it came from, then feed sugar to the hives instead
- in the same way, do not use pollen of unknown origin
- place hives in an arrangement that minimises drift of bees between hives, such as a semicircle, square or 'snake' pattern
- do not take honey, bees, or brood from one hive and spread it around several others unless you are very sure that the source colony is free of AFB.

Regulations concerning AFB disease control

The location of all apiaries must be registered if beehives are kept there for more than 30 consecutive days. This applies whether the beekeeper has only one hive or thousands.

Beekeepers who find AFB in hives must kill the bees and brood, notify the PMS management agency in writing within seven days, and take further steps as approved in their disease elimination conformity agreement (DECA). Beekeepers without a DECA must destroy the bees and all hive parts by fire.

All hives must be inspected at least once a year by an approved beekeeper. An approved beekeeper is someone who has passed a competency test on AFB recognition and control, and has a DECA approved by the PMS management agency. Hive owners who are not approved beekeepers must obtain the services of one, who will inspect and report on the hives by 1 December each year, and complete a certificate of inspection.

Approved beekeepers do not have to complete a certificate of inspection and can inspect and report on their own hives. Reporting is usually done as part of the annual disease return.

Every registered beekeeper is required to complete a hive inspection statement, called an annual disease return or ADR (which is provided by the PMS management agency), and return it to a nominated AsureQuality office by 1 June each year. Alternatively, beekeepers can print their own ADR from ApiWeb (see below) and post it to AsureQuality.

The combination of information about hive location and disease incidence is essential for obtaining an overall pattern of disease outbreaks and apiary locations. Apiary locations and contact details for beekeepers are kept in an apiary register maintained by AsureQuality Limited on behalf of the PMS management agency and MAF Biosecurity. Beekeepers can now access and amend their own address and apiary details on-line using the password-protected website ApiWeb (https://apiweb.asurequality.com).

No drugs may be used for the prevention or treatment of any bee disease in New Zealand except those approved for the purpose. None has been approved for preventing or controlling AFB.

Sacbrood

Sacbrood is not usually serious, but beekeepers must be familiar with its symptoms as they are frequently confused with those of AFB and the two diseases can occur together. Sacbrood is caused by a virus called sacbrood virus.

Symptoms: colony

Seriously infected colonies display a patchy brood pattern with sunken cappings, which in many cases may be perforated. This brood pattern may appear similar to combs containing brood infected with AFB.

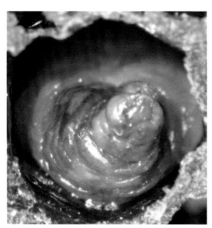

Fig. 13.25 Larva infected with sacbrood, showing the darker head.

Symptoms: brood

Death occurs after cells have been capped, and when the larva is lying flat along the bottom of the cell. An infected larva turns yellow, grey, and later almost black. The head darkens rapidly before the remainder of the body changes colour. This is an important characteristic (Fig. 13.25).

The skin of the dead larva changes into a tough, plastic-like sac, which gives the disease its name. Inside this sac the body contents of the dead larva are very watery (Fig. 13.26). Infected larvae maintain their body segmentation, unlike larvae that have died from AFB, which lose this segmentation and become smooth-sided.

Larvae that have died from sacbrood can usually be lifted out of the cell with a matchstick, but if the sac is punctured the body contents will run out. The contents will not rope with the smooth, even-coloured brown rope characteristic of AFB. If the contents rope at all, the thread will not be elastic and will be lumpy and uneven in colour.

Larvae that have died from sacbrood can easily be removed from the cell, even after a long time when they have dried out to a type of scale. There may be a putrid odour in some advanced stages.

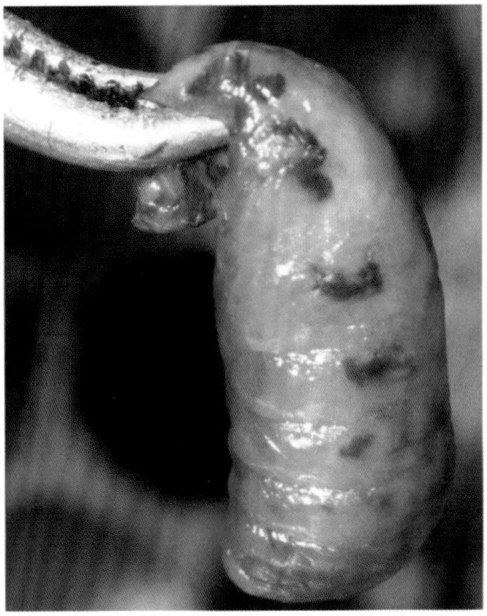

Fig.13.26 Larva infected with sacbrood, showing fluid under skin.

Transmission of the disease

Sacbrood virus is present in many apparently healthy colonies, but sacbrood disease becomes noticeable only in some circumstances. Young larvae are infected with the virus when it is ingested along with food, and the virus multiplies and attacks the tissue of the larvae. In light infections, hive bees remove dead larvae from the cells so efficiently that the disease is not noticeable during inspection.

Treatment

Sacbrood disease is not normally a serious problem, and infected colonies may recover of their own accord. The disease is most common in spring, and it generally does not persist over the honey flow. In severe or persistent cases the only effective treatment is to requeen the colony with a young queen of a good strain, or to requeen and strengthen the colony simultaneously by uniting a nucleus onto it.

Sacbrood can be confused with American foulbrood. If in doubt, contact an experienced approved beekeeper, an authorised person, the PMS management agency or an apiculture officer of AsureQuality.

Chalkbrood

Cause

Chalkbrood is caused by the fungus *Ascosphaera apis*. Honey bee larvae eat spores of this fungus with their food, with larvae three or four days old being the most susceptible to infection. The spores germinate in the larval gut, after which the fungus grows inside the gut as vegetative filaments or hyphae.

The mass of hyphae (sometimes called the mycelium) breaks through the gut wall and invades the larva's body, though the head is often left intact. Spore formation may take place later in fruiting bodies on the outside of the infected larva.

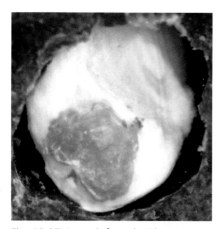

Fig. 13.27 Larva infected with chalkbrood.

Fig. 13.28 Chalkbrood mummies at the hive entrance, some with dark fruiting bodies.

Symptoms

Larvae almost always die at the elongated stage, and only rarely when coiled in the bottom of the cell. Both capped and uncapped cells can contain diseased larvae, so it will sometimes be necessary to pick the cappings off any odd-looking cells to diagnose the disease, in the same way as when checking for AFB.

Larvae infected with chalkbrood first turn vivid white, much whiter than the normal pearly-white of healthy brood. At this stage the body has a furry surface because of the fungal mycelium, especially towards the rear of the cell. The head is usually still visible (Fig. 13.27).

The larval body later dries out and becomes hard, and at this stage the remains are termed mummies. They are mostly white, though some are partly or completely grey to black because of the growth of fruiting bodies on the mummy surface. A comb containing a lot of mummies will rattle when shaken.

Larvae infected with chalkbrood can be mistaken for mouldy or white pollen. Examination of the cell contents shows the difference: pollen is packed in the cell as a solid mass, whereas a soft chalkbrood-infected larva retains some of the shape of a honey bee larva. Infected larvae can easily be removed from the cell.

Mummies are easily removed by bees from the cells, so can often be seen on the floorboard or on the ground outside the hive entrance where the bees have deposited them (Fig. 13.28).

Chalkbrood and AFB can be in the same hive (Fig. 13.29), and if chalkbrood infection is heavy, AFB larvae may not rope out as much as in hives that do not have chalkbrood. The chalkbrood fungus is believed to have an antibiotic effect on the AFB, and the rope may only extend 10–15 mm instead of up to 30 mm or more.

Spread

Chalkbrood is thought to be spread mostly by the beekeeper, for example by exchanging equipment and bees between colonies, and feeding pollen or honey. The other main cause of disease spread in apiaries is probably through drifting of adult bees between hives.

Spores of the chalkbrood fungus remain viable for at least 15 years, and in areas where chalkbrood is present there seems to be a low level of the disease in many apiaries. Weak hives and nuclei may have more chalkbrood than strong hives and a shortage of good-quality pollen or protein may allow more chalkbrood to develop.

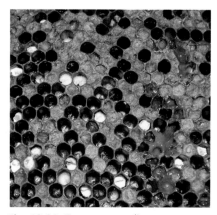

Fig. 13.29 Two or more diseases can be found in the same hive: AFB and chalkbrood together.

Control

Chalkbrood rarely kills a colony, although heavy infections will reduce the worker population significantly. The disease can occur at any time, but is most common during spring and symptoms often disappear with the onset of the honey flow.

Chalkbrood mummies can be easily removed from the comb by hive bees (unlike AFB scale), and so colonies are usually able to keep infection at a low level. Control of chalkbrood is rarely necessary.

Prevention

Sound management practices can reduce the impact of chalkbrood in an apiary.

Bee colonies The most important prevention measures are to use young, productive queens to keep colonies strong at all times, and ensure the colonies have access to good-quality pollen or are fed pollen substitute.

Apiary location Drifting is probably a cause of spread between hives, so position hives in a low-drift layout (see Chapter 2).

Bee stock It may be possible to increase disease resistance by breeding from queens whose colonies show a low degree of chalkbrood infection.

European foulbrood

European foulbrood (EFB) is common in many parts of the world, but has not been found in New Zealand. It is a notifiable organism, which means that anyone who suspects EFB in New Zealand must report it immediately — to MAF Biosecurity on 0800 809966 or to an AsureQuality apicultural officer. EFB has been suspected occasionally in New Zealand, but extensive surveying and testing have shown the disease not to be present.

In Europe and North America, EFB is regarded as a minor disease and is controlled by using resistant bee stocks and antibiotics. EFB has had a much more significant impact in Australia. The effects that EFB would have in New Zealand are unknown, but could be far-reaching. Beekeepers should be aware of its symptoms in case it is inadvertently established. EFB could enter New Zealand in bee products, on bees, combs of brood and honey, or on used beekeeping equipment.

EFB is caused by the bacterium *Melissococcus plutonius*. Honey bee larvae ingest this bacterium with their food, in much the same way as with the AFB bacterium *Paenibacillus larvae larvae*, and it multiplies inside the gut where it competes for food. The EFB bacterium does not form spores as the AFB bacterium does, but it is resilient and can survive for many years.

Larvae infected with EFB may starve, in which case they usually die when four to five days old. Surviving larvae may be undernourished and emerge as smaller adults.

The disease can be present in a colony without any signs being visible, but changes in the ratio of bees to brood, the weather and in food intake may result in symptoms appearing. This typically occurs in the spring and disease symptoms may disappear over the honey flow.

Symptoms: colony

As with AFB, the foul odour is not always present — brood combs infected with EFB may smell sour (rather like wet nappies) or may not be smelly, depending on what species of secondary bacteria have grown in the larval remains after death.

In advanced cases of EFB, brood combs will have a patchy pattern. Most larvae die before being capped, but any cappings over infected larvae will be dark, sunken and perforated. AFB larvae, on the other hand, mostly die after the cells are capped.

Symptoms: brood

Most larvae die when they are coiled in the base of the cell in the 'C' shape, but some may die during the transition to the pre-pupal stage when they are stretching out along the cell (Fig. 13.30). Larvae in colonies affected by parasitic mite syndrome or half-moon syndrome also die during the larval or pre-pupal stage and can spiral up the cell wall. AFB larvae always die in the pre-pupal to pupal stage, and never coiled in the bottom of the cell.

With EFB look for off-colour larvae in cells not capped over. Sometimes the cells will be capped and become sunken with small holes torn in them as happens with AFB, but most will not be capped.

After larvae die from EFB they darken, turning yellow, light brown, dark brown and sometimes almost black. The tracheae (breathing tubes) appear as white strands in the remains and are often quite prominent. AFB pre-pupae go light brown to dark brown, and do not show prominent tracheae.

Fig. 13.30 EFB-affected larva in C-shape and half-moon positions.

With EFB, the larval remains are first watery,

then become glutinous, and finally dry out to a rubbery scale. Sometimes dead larvae will rope out in an uneven thread, up to 15 mm long, but the rope is not as even or as long as with AFB. PMS-affected larval remains, on the other hand, will be white to yellow in colour but can be lifted out of the cell.

Other brood disorders

Honey bee larvae can die from many different causes, and an experienced beekeeper should be able to determine whether brood has died from AFB, chalkbrood or sacbrood, some other disorder or syndrome, or other causes such as chilling, overheating, pesticide poisoning, starvation, non-viable eggs or inbreeding.

Half-moon syndrome (HMS)

This brood condition has been called half-moon syndrome (or disorder) from the shape adopted by some dead larvae. Half-moon syndrome does not appear to be a disease, as no causative organism has been found. It is thought to be a condition related to an abnormality of queens caused by poor nutrition during their early adult life.

The most obvious symptoms are:
- Many brood cells contain multiple eggs, often laid on the walls of cells. Eggs are sometimes glued together end-on-end like sausages. Multiple eggs are a feature of laying workers, but eggs from laying workers are scattered all over the cells and usually in singles. Multiple eggs are not a feature of EFB, PMS or AFB.
- Capped brood may be perforated in advanced stages.
- There is usually a lot of scattered drone brood in worker cells.
- Larvae die when uncapped and twist up the cell wall as seen with EFB and PMS. Larvae do not usually die in the base of the cell.
- The larvae turn yellow, light brown then dark brown, and often coil around the walls and mouth of the cell. Ultimately the remains dry out to a scale that may lie in a half-moon or C-shape, often at the lip of the cell.
- The scale can be removed easily, unlike that caused by AFB.
- Dead larvae are not ropy, and their tracheae may become prominent.
- Sometimes there are supersedure cells, which may or may not show HMS symptoms.
- Requeening usually corrects the problem.
- The brood has a musky urine-like smell, not an acrid smell like EFB.

Larval remains affected by HMS can look very similar to EFB-affected larvae, so it is always a good idea to contact an AsureQuality apicultural officer for a second opinion, and so arrangements can be made to collect a sample for testing for EFB.

Table 13.2   Characteristics of American foulbrood (AFB), sacbrood, chalkbrood, European foulbrood (EFB), parasitic mite syndrome (PMS) and half-moon syndrome (HMS)

Feature	AFB	Sacbrood	Chalkbrood
Sealed brood pattern	Patchy in advanced cases	Patchy in advanced cases	May become patchy
Cappings	Sunken, darker in colour, perforated	Often removed, can be sunken and perforated	Become dark and sunken in heavy infections, perforated. Affected pupae clearly visible in uncapped cells
Age of dead brood	Young pupae or pre-pupae. Often die after capping	Pre-pupae	Pre-pupae
Colour of dead brood	Off-white then light brown then dark coffee brown to black, coloration even	Off-white then brown, to grey, becoming black, head end usually darker than body	Larvae are fluffy white, with a yellow head, dry to hard chalky mummies either creamy-white or grey-black
Segmentation	Loses early	Retains	Visible
Consistency of dead brood	Smooth rope, 10–30 mm long when at brown stage, later tacky	Plastic sac with watery contents (often lumpy)	Initially soft with a furry surface swollen to a hexagonal shape filling the cell, later shrinking to a hard, chalky mummy
Pupal mouth parts	Prominent in dead pupae, may be attached to opposite wall of cell	Rarely prominent, never reach opposite wall of cell	Not noticeable, but head is a conspicuous dark spot in soft larvae
Brood remains	Dark scales, sometimes tacky, which stick tightly to bottom of cell	Dark scales, easily removed from cell	Creamy-white or grey-black mummies easily removed from cell. Also found on floorboard or at hive entrance
Smell	Putrid in advanced cases	Sour smell present occasionally	Not noticeable

EFB	PMS	HMS
Irregular	Patchy in advanced cases	Drone brood in worker cells. Multiple eggs often laid end to end and on cell walls. Patchy in advanced cases
Infected larvae mostly in uncapped cells. Some cappings discoloured, sunken, perforated	Some sunken, darker in colour, perforated	Some cappings discoloured, sunken, perforated
Mostly larvae before capping, usually in coiled stage	Varies from young C-shape larvae to pre-pupae, not pupae	Mostly larvae before capping, usually in coiled stage
Dull white, then yellowish, often turning brown or almost black, white distinctive tracheae	White to yellow then grey to light or dark brown in scale stage	Off-white to yellow with prominent tracheae then dark brown
Visible	Not very distinct	Visible
At first watery, then pasty, not usually ropy. Scales tough	Can remove and not ropy. Contents fragile	Can remove and not ropy
Brood dies before pupal stage	Brood dies before pupal stage	Brood dies before pupal stage
Very irregular position often twisted up cell walls or sometimes extended along cell. Tracheae often prominent, scales rubbery and easily removed	Slumped along lower wall or may twist up cell walls and dry to a soft scale. Varroa often present	Slumped along lower wall or may twist up cell walls. May dry to a soft scale up cell wall or around the lip of the cell like a half-moon
Usually none in early stages, sour acrid smell may be present later	May be slightly sour smell	Not pronounced, musky to urine-like

CAUSES OF ADULT BEE MORTALITY

The symptoms of adult bee diseases are difficult to diagnose, and their presence is complicated by other factors such as chilling and poisoning. At certain times of the year there will be more than 1000 bees per colony dying each day from natural causes unrelated to pests or diseases. Although insect-eating animals such as hedgehogs and birds remove a lot of these dead bees, some will inevitably collect at the hive entrance. Unless there are more than one or two cupfuls of dead bees present, there is probably no cause for alarm.

Nosema disease

Nosema disease is caused by infection with different species of the microsporidian genus *Nosema*, the most well known being *Nosema apis*. *Nosema ceranae* was found in both the North and South Islands of New Zealand in 2010, but its distribution and impact on honey bee colonies are unknown.

Nosema apis spores are consumed by bees when they clean infected comb and remove dead bees. The spores germinate in the bees' digestive tract and infect the gut lining. There they multiply rapidly and produce more spores, which are passed out of the bee with the faeces. The spores are very resistant to dehydration.

Nosema apis is present in almost all colonies in New Zealand throughout the year, but is more common in spring and winter. Spore counts of nosema increase when a colony is short of food, confined to the hive for a long period, or when brood combs are not renewed on a regular basis. Nosema levels need to average over 10 million spores per bee before the effects of the disease are noticeable, though the disease will no doubt be having some effect at lower levels of infection.

The disease is more important in queen-raising colonies because nosema infection of queens greatly increases their supersedure rate. Queen breeders exporting their product try to provide nosema-free bees for cage escorts.

The only certain way to diagnose this disease is to use a microscope at 400x magnification, though this is not sufficient to distinguish between the different *Nosema* species.

The product Fumidil B (with active ingredient fumagillin) is approved for controlling nosema disease. Fumidil B is a white powder that is mixed in sugar syrup and fed to bees to prevent nosema spores from germinating in the bee's stomach. As it has no effect on the spores in the resting stage, Fumidil B is usually fed for at least four weeks, so that the nosema is killed as it germinates from the spores. The product's use is restricted to hives that are used for rearing, mating or storing queen bees. It must not be used in hives producing honey that is likely to be used for human consumption.

Paralysis

There are at least 20 viruses currently known to affect honey bees, and some of these have been found in New Zealand. Paralysis is caused by several different viruses, which may be present in a colony without producing any noticeable symptoms. In New Zealand higher levels of paralysis infection will result in some adult bees becoming hairless and black with stunted and somewhat pointed abdomens. They move in a trembling motion with wings outstretched, and are often seen being evicted from the hive by guard bees.

Occasionally a lot of dead and dying bees will accumulate outside a hive entrance, with trembling bees walking about over a pile of dead bees. If insecticide poisoning is not likely to be the cause, the bees could be affected with paralysis. Requeening is usually effective in clearing up this problem.

Dysentery

Dysentery is not a disease, but a disturbance of the digestive system that causes bees to defecate on the combs or near the hive entrance. It may be caused by fermented honey stores, excessive moisture in the hive or confinement to the hive by bad weather.

Dysentery may lead to the spread of nosema disease. Dysentery can be alleviated by making sure the hive is dry inside, and that the colony has adequate stores of capped honey.

Insecticide poisoning

Bees foraging on flowers that have been sprayed with an insecticide will often die away from the hive, but a significant quantity (say two or three cupfuls) of dead bees may build up inside and outside the hive, suggesting a problem.

Insecticide poisoning happens particularly in fruit-growing areas, and also in towns and cities as a result of the careless use of chemicals in home gardens. Tracing insecticide damage to a particular source is difficult and proving liability even more so. Chapter 20 covers legislation related to pesticide use.

Tanalised timber

Tanalised timber contains arsenic compounds that leach out of the wood, so generally it cannot be used for beekeeping equipment. Exceptions to this rule are hive parts that don't come in contact with the bees, such as ground runners and rims of hive lids.

Arsenic poisoning from tanalised hive parts is rarely dramatic, but usually acts as an insidious brake on colony development, especially in spring. Beekeepers who claim to use tanalised timber successfully may simply not be noticing these subtle effects.

Chilling

Bees foraging for nectar and pollen may become chilled and die outside the hive entrance, especially in spring. This is more obvious when bees laden with pollen are found crawling on the ground in front of hives up to several metres away.

Toxic and narcotic nectar

The native tree karaka (*Corynocarpus laevigatus*) produces nectar that is highly attractive, yet very toxic, to bees. The nectar is not known to be toxic to humans, although the berries are unless cooked correctly. Karaka has large, glossy, dark green leaves, and conspicuous yellow to orange berries in late summer or autumn. It is common in coastal areas in the north of the North Island, is also found in the South Island, and is used as an ornamental tree.

Fig. 13.31 Lesser (L) and greater (R) wax moth adults.

Karaka flowers in October, and if trees are within flight range of your bees and few other nectar sources are available, you should consider moving your colonies away during the flowering period.

Symptoms of karaka poisoning are similar to those of pesticide poisoning, with bees found crawling on the ground in front of the hive. In serious cases the adult population of the hive will become depleted, leaving behind only a small handful of bees and the queen. The only way to positively identify karaka poisoning is to analyse the dead bees' stomach contents under the microscope for the presence of karaka pollen grains.

Kowhai species (mainly *Sophora tetraptera* and *Sophora microphylla*) produce nectar that may have a narcotic effect on bees in some years. Affected bees may stay out in cold, wet weather and become too chilled to return to the hive. The narcotic effect of kowhai nectar rarely affects colonies seriously, and kowhai is actually a valuable spring nectar source in many areas.

Colony collapse disorder (CCD)

Colony collapse disorder (CCD) is one of many terms used down the years to describe unexplained, large-scale loss of honey bee colonies. This particular one was coined in the USA in the mid-2000s to describe a sudden population loss in a colony with few, if any, associated dead bees in front of or inside the hive. Beekeepers observed that either all the bees left the hive, or just the queen and a few workers remained. Another characteristic was that, before absconding, a populous colony with CCD showed a loss of cohesion with worker bees scattered all over the brood. Similar observations were made in some European countries.

As with other large-scale colony losses, there are many theories about the causes of CCD. These include infections with *Nosema ceranae* or viruses (especially acute Israeli paralysis virus and deformed wing virus), poisoning by a class of pesticides called neonicotinoids, the use of miticides, mite infestation and poor nutrition. It is likely that the actual cause is a complex interaction between several pathogens and other environmental conditions.

Colony collapse disorder has not been reported in New Zealand. Beekeepers have seen colonies absconding and losing adult bees, often with the honey crop on the hives, though advanced PMS symptoms are usually present in such hives with patchy brood as well as plenty of varroa mites.

If you think you are seeing large-scale CCD, contact an AsureQuality apicultural officer or the MAF Biosecurity hotline 0800 809966.

BEEKEEPING PESTS

Wax moths

Wax moths are a serious pest of the beekeeping industry, damaging bee combs, comb honey and other bee products. There are two wax moth species in New Zealand: the greater wax moth *Galleria mellonella* and the lesser wax moth *Achroia grisella* (Fig. 13.31). They are not native, but were probably imported accidentally with bees and beehives. Other stored-product moths may also occasionally infest bee combs.

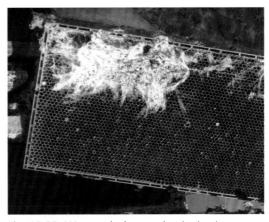

Fig. 13.32 Wax moth damage beginning in one corner of a comb.

The distribution of wax moth species is controlled mainly by climate. Serious damage is caused by the greater wax moth in most of the North Island and warmer parts of the South Island. The lesser wax moth appears to have more tolerance to cold and is found throughout New Zealand.

The greater wax moth is by far the more destructive, but both species have similar habits and the same methods are used to control them.

Life cycle

Female moths lay eggs on bee combs or in cracks in hive woodware. These eggs are small, white and very difficult to see. They hatch in about a week outside the hive, but in a few days in the warmer conditions of the hive.

Fig. 13.33 Combs completely destroyed by wax moths.

Larvae are white and about 1 mm long when they first emerge. They take their first meal of honey or pollen and then burrow from the surface of the comb down to the midrib. There they tunnel through the comb, feeding on honey, pollen, beeswax and the general debris that remains in the comb after brood rearing. Their tunnels are lined with silk and spotted with faecal pellets (Fig. 13.32). Damage to the comb is rapid, and it soon becomes criss-crossed with silken tunnels. Ultimately the entire substance of the comb is reduced to a mass of frass and debris (Fig. 13.33).

The time wax moths spend in the larval stage varies considerably. In cold conditions or when there is a shortage of food it may be as long as six months, but in warm conditions

Fig. 13.34 Pupal cocoons of the greater wax moth that have filled the gap between top bars and the hive lid.

and with abundant food larvae may become fully grown in as little as four weeks. In these conditions the damage to combs is greatest.

When fully grown, greater wax moth larvae are up to 30 mm long. At this stage they move to a wooden part of the hive to pupate. Bee spaces, especially at the end or top bars of frames, are most often chosen (Fig. 13.34). Before they pupate, greater wax moth larvae gnaw a shallow depression in the wood. A large number of these can seriously weaken frame end bars.

Pupation takes place in a silken cocoon, with many cocoons clustered together. The pupal stage of the greater wax moth lasts about 10 days in warm conditions.

Lesser wax moth larvae also pupate in silk cocoons, but singly rather than in the large congregations of the greater wax moth. Pupal cocoons may be found throughout the hive, especially in sacking hive mats. Pupation takes about 16 days in warm conditions.

Where wax moth causes damage, and how to control it

Hives Wax moths are never a problem in beehives that are inhabited by a strong colony of bees. Most colonies tolerate a low population of wax moth, but will not let numbers increase to a point where much damage occurs.

If a bee colony is weakened or dies out through queen failure, starvation, disease or insecticide poisoning, wax moths will be free to multiply and infest the hive. In warm weather the combs in an empty hive may be destroyed in as little as two or three weeks.

The best defence is to ensure that all colonies remain strong. Any hives that die out should be restocked with bees, or taken into storage and dealt with in one of the ways outlined below. But first ensure that the cause of death was not American foulbrood.

Stored comb Stored combs are not protected by the bees, so may be damaged by wax moths, particularly in warmer parts of the country. Combs culled because of damage, excess drone cells, or pollen clogging should be processed without delay, as wax moths will readily infest them.

Because of the feeding habits of wax moths it is good practice to sort combs into three categories to separate areas of risk and contain any infestations. The categories are:
- clean, 'white' combs in which no brood has been reared and no pollen is stored
- combs that have been used for brood rearing, or which contain pollen
- cull-combs waiting to be melted down.

To store combs, first space them evenly in each super. Then put the supers in stacks with division boards or hive lids forming a seal top and bottom. Put queen excluders in the stacks every three or four supers to minimise damage should rats or mice gain entry. Set approved rodent bait stations or traps around the area where combs are stored. Use masking tape to seal any cracks or holes in supers that could admit moths.

There are three recommended methods of protecting stored combs, and we will mention two others.

Freezing

Placing honey combs (and supers) in a freezer for 48 hours will kill all stages of wax moth present, including eggs. The supers should then be stored in stacks, sealed off top and bottom as described above. Supers may also be stored in plastic bags. If care is taken to prevent reinfestation, freezing is an easy and effective method for controlling wax moth. For large quantities of honey supers or honey combs it will take longer than 48 hours for the core temperature to reach freezing, as honey and wax are very good insulators.

Cooling

In areas with cold winters, sort combs as described and store them in a shed that is as cold and draughty as possible. Some commercial beekeepers in these areas build comb barns with slatted floors and extra ventilation to discourage wax moth infestation. An alternative to a slatted floor is to lift stacks of supers off the floor on strips of timber. In either case, the tops and bottoms of stacks should be covered with queen excluders or mesh screens to admit plenty of air but exclude rodents. Some hobby beekeepers with only a few supers take the combs out of the supers and hang them on racks where they are exposed to cool or cold air and maximum light, both of which discourage wax moth development.

Some commercial beekeepers keep sheds or insulated shipping containers cool using air-conditioning units. This protects combs over the vulnerable summer and early autumn months. Other commercial beekeepers find it useful to store honey supers as 'wets', without drying them out, and sometimes put sheets of newspaper between supers to limit the spread of the moths.

Storage on hives

Supers of empty combs can be protected by leaving them on hives over winter. Put them over a half-mat, or an escape board with the bee escapes removed, to allow limited bee access. As bees do not heat the air around them in order to stay warm, they do not have to heat this extra space.

Fumigation

There are few, if any, methods available for safely fumigating combs. Wax moth crystals or paradichlorobenzene (PDB) used to be sold to control wax moths, but the chemical is not registered for this use and has caused residue problems in export honey and propolis in recent years. Moth ball crystals (naphthalene) are similar to PDB in appearance and odour, and are readily absorbed by wax. Do not use PDB or mothball crystals on stored bee combs.

Do not use 'borer bombs' or any aerosol insecticide sprays in or near comb storage areas, as the insecticide is absorbed by the wax and kills bees when the combs are placed on hives. Methyl bromide and ethylene oxide used to be used on stored combs and comb honey, but are no longer a practical option.

Traps

Some beekeepers make traps for comb storage rooms from plastic soft-drink bottles. Drill a 20–25 mm hole just below the shoulder of the bottle and add a mixture of one cup of water, one cup of sugar (any type), half a cup of white vinegar and a banana peel. This will ferment and is supposed to attract and trap wax moths. Pheromone traps for male wax moths are available overseas.

Insect traps using UV light with electrocutors, water baths or sticky glue boards may have some effect, but the damage to combs is really done by the larvae that hatch from eggs already present in the combs when they come into storage. This means that traps of any sort are of limited value.

Comb honey for human consumption Although comb honey will seldom have active larval infestation when packed, eggs may be present. After packaging, and when placed in a warm place such as a kitchen cupboard or shop shelf, these eggs may hatch and the larvae will damage the product.

Comb honey for human consumption must not be treated with a chemical fumigant as the honey will be tainted. The best method of treating comb honey is to freeze it to kill wax moths, using the same method as for empty combs. The honey must then either be processed immediately, or stored in moth-proof stacks with newspaper between supers to prevent reinfestation. Large-scale comb honey producers use commercial cool stores to freeze their honey.

Feed honey Frames of honey stored at the end of the season for bee feed in the next spring may become infested by wax moth. To help prevent this, feed honey can be left on the hives until the first frosts before being stored, so the cold conditions can inhibit wax moth development.

Unless the area where the honey is stored off the hive is very cold, wax moth infestation may develop. The only approved treatment for controlling wax moths in feed honey is freezing. The honey can be stored at room temperature after freezing provided the supers are moth-proof. Freezing allows the honey to be harvested earlier if varroa treatments need to be applied in the summer months.

Fig. 13.35 Common wasp attacking a worker honey bee.

Pollen Bee-collected pollen is very readily infested by wax moths, and this often starts in the pollen trap before harvesting. If untreated the pollen may be completely destroyed.

To prevent wax moth damage, harvest pollen from the traps at least twice a week, then place it in a freezer for 48 hours or until needed. Pollen can also be fumigated with food-grade carbon dioxide by putting it in a plastic bag and displacing the air with the gas.

German and common wasps

Four species of social wasps have established in New Zealand after being accidentally introduced from other countries. Two are paper wasps that have no effect on beekeeping, but the other two species are a more serious threat.

The German wasp *Vespula germanica* was first identified in New Zealand in 1944, and is found throughout the country. A very similar wasp, the common wasp or *Vespula vulgaris* (Fig. 13.35), became established in about 1980 and has colonised most parts of New Zealand. German and common wasps are very similar in appearance and life history, and the same control measures are used for both species.

Wasp populations seem to wax and wane over a period of years with corresponding changes to their effects on beehives, and the reason for this is not known. Species of the parasitoid wasp *Sphecophaga* have been released in New Zealand since 1987, but have had limited success in controlling wasp numbers.

Both the common wasp and the German wasp can be identified by their conspicuous yellow and black body colours. They are slightly bigger than a honey bee and have a smooth, rather than hairy, body. Both species have a black head and thorax, with yellow and black stripes on the abdomen.

Wasps are a particular problem because their nests become strong in autumn when honey bee colonies are diminishing in population. Wasps are also active earlier and later in the day than bees, and in worse weather conditions. Wasps most frequently gain access to a hive when the bees are clustering inside the hive and are unable to guard the full width of the entrance.

To help bee colonies guard against wasp attacks, put entrance reducers on hives when preparing the hives for winter. If a very weak hive is under severe wasp attack, reduce the entrance to about 12 mm wide or move the hive to another locality.

Finding nests

Very high wasp numbers indicate a nest (or nests) within several hundred metres. Efforts should be made to find these nests and destroy them if the terrain allows, as large numbers of wasps can seriously weaken the colonies in an apiary.

Flying wasps can be tracked in early morning or at dusk when other insects such as bees and flies have stopped flying. To make them more conspicuous, dust the wasps with flour or icing sugar. First catch some wasps on hive fronts or off combs, using a glass or plastic jar that is dusted inside with flour or icing sugar. Roll the jar around a few times to ensure the wasps are well coated, and then release them to fly back to their nests. It may be possible to get a 'wasp-line' at this time and follow the flight path to nest locations.

Destroying nests

The most effective method of destroying a wasp nest depends on its situation. Wasps can be dangerous, so take the following precautions:

- carry out the work in the evening, after the wasps have stopped flying
- wear full protective clothing, including gloves and a veil.

If in any doubt over your personal safety and that of others nearby, engage the services of a professional pest exterminator. If using insecticides always wear appropriate clothing and gloves and follow label directions.

Nests in the ground Flying wasps entering and leaving a small hole in the ground usually indicate that a nest lies underneath. Powdered insecticides generally work well, as the wasps carry the powder into the centre of the nest. Carbaryl powder is most effective and is available in garden and hardware shops. Dust around and into the entrance of the nest after dark when all the wasps have returned for the night (Fig. 13.36). Use 5–10 g (about 1–2 teaspoons) of the insecticide powder, and do not plug the hole. Very large nests may have more than one entrance, so try to find and dust them all.

Some insecticides, especially carbaryl, degrade over time if stored in warm conditions, so check a day or so after applying to see if the insecticide has been effective.

It may take a few days to kill all the wasps by this method, as young ones will continue to emerge from the brood combs. In rare cases it is necessary to apply more insecticide after about seven days. When all adult wasps have died, plug the hole to prevent new wasps from emerging.

Fig. 13.36 Dusting insecticide powder at a wasp nest entrance.

Nests in roofs and basements Where nests are readily accessible, sprinkle powdered insecticide around the entrance(s) as for nests in the ground, or try spraying the nest directly with insecticide solution. The advantage of powdered insecticide is that the wasps carry it into their nests on their bodies, and spread it around as they groom each other and feed the larvae.

If the nest is not easily accessible, it may be possible to tape or screw a small can onto the end of a long pole and place the insecticide powder in that. Use the can to break through the surface of the nest and deposit the powder into the nest. Wasps will emerge from this hole so be prepared for a quick retreat.

Nests in walls If the wasps are entering a wall through a small hole, make sure that alternative entrances are blocked and then use one of the following methods to destroy the nest:

- Cut off about 100 mm from a hose that fits into the entrance hole. Pack a small quantity of powdered insecticide into it and tap the hose until the powder has compacted to one side. Place the hose in the hole so that the wasps walk over the insecticide to gain access to their nest. They will track the insecticide in and spread it around the nest. Repeat several times. It may pay to let the wasps get used to their new restricted entrance before applying the insecticide.
- Blow insecticide in. Buy insecticide powder in a puffer pack or put insecticide powder into a plastic squeeze bottle with a small nozzle. Puff insecticide into and around the entrance to the hole. Wash out the bottle after the application.

Poison bait

There is no reliable method of controlling wasps with poison baits, but this method can be attempted in an apiary if wasps are numerous. Use synthetic pyrethroid insecticides, as these tend to be less repellent than compounds such as carbaryl.

Spray the insecticide solution onto protein bait such as meat or fish, and put the bait out for the wasps. Always put the bait in a bait station such as an empty soft drink tin, located where it cannot be interfered with by children or household pets. Put out only enough bait for two or three days' feeding. It may be useful to put out protein baits first without the insecticide, to see if wasps can be attracted to feed on the baits, before applying the insecticide.

If wasps are feeding on carbohydrates (sweet food) to the exclusion of protein, do not be tempted to mix insecticide with sugar or honey as honey bees will be attracted to the bait and poisoned.

This method of using poison baits gives mixed results mainly because, unlike the honey bee, each wasp forages individually and does not communicate the location of a food source. The lack of an insecticide that does not repel wasps is also a problem.

Dusting insecticide on foraging wasps

Some beekeepers report success by dusting wasps that are robbing beehives, using the method described above for tracking wasps but with insecticide rather than flour or icing sugar. This is a slow process. Continued applications are needed, especially if the wasp predation is heavy, as the wasps will probably be from several different nests or a very large nest. Do not dust the wasps on the hive or combs, as this will leave a toxic residue that will kill honey bees.

An easier way to do this is to use a dedicated battery-powered vacuum collector to suck up wasps. When you have a number of wasps inside the collector, suck up a teaspoon or so of carbaryl powder. Shake the collector to distribute the powder on the wasps, then unscrew the collector and release the wasps.

Wasp traps

Traps can be made or purchased but beekeepers do not report having great success with them.

Shifting hives

If destruction of nests in the immediate vicinity of the apiary does not reduce wasp numbers, remove the hives to another site for winter. The hives need to be shifted at least three kilometres to minimise the number of bees returning to the apiary site.

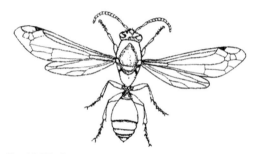

Fig 13.37 Paper wasp.

Paper wasps

The Australian paper wasp *Polistes humilis* (Fig. 13.37) is found mostly in the northern half of the North Island. The wasps are brown and delicate in appearance, and build nests of wasp paper that hang by a single stalk from branches or the eaves of buildings and can grow to about the size of a saucer.

A related species, *Polistes chinensis*, is found throughout the North Island and the top of the South Island, and is also established in Central Otago and Dunedin. It has the same delicate body as the Australian paper wasp, but is black and yellow in colour so is sometimes confused with the German wasp. It builds a similar nest to the Australian paper wasp, but tends to build in bushes, hedges, on wooden fences and under eaves.

Paper wasp control

Neither species of paper wasp have any effect on beekeeping. If nests are isolated they can be left alone, although these can be a nuisance as paper wasps have fierce stings. People can be surprised by them when bumping or pruning bushes.

If you do wish to destroy exposed paper wasp nests, thoroughly wet the nests with an aerosol insecticide in the evening when all the wasps will be clustered on the outside of the nest. Sprayed wasps will fall to the ground and may take some time to die, so wear shoes or boots and stay away from the nest area after spraying until all the wasps are dead.

Mice

In autumn mice often choose beehives for dry winter quarters. They can make a mess by chewing out several combs in the back of the hive (Fig. 13.38) and bringing in quantities of straw. Mice can get through any holes more than 10 mm across, so in autumn and winter make sure that hive entrances are reduced to 8 or 9 mm high and that there are no cracks in the floorboard or boxes. Mice often burrow into the dry ground under a hive, so in autumn it may pay to lay rodent baits under several hives in each apiary.

Fig 13.38 Mouse nest damage in combs.

14 Harvesting honey

As a beekeeper you are privileged to be able to enjoy fresh honey produced by your own bees. Taking the honey off the hives can be hot and back-breaking work, but is one of beekeeping's greatest satisfactions.

WHEN TO HARVEST

Honey can be harvested as the season progresses, or all at once when honey production stops. If you choose the latter option you will need to set up and clean an extraction plant only once each season, and you will need to visit your apiaries less often.

However, as a general rule progressive honey harvesting is a much better beekeeping practice because:

- fewer supers are needed, as you can use them more than once during the honey flow
- putting freshly extracted supers onto a hive during the honey flow seems to stimulate the bees to store more honey
- honey is easier to extract from the comb in warmer weather and while still warm from the hive
- there is less likelihood of robbing when honey is harvested during the honey flow
- the workload of extracting is spread out, leaving you more time in late summer/autumn for such things as queen rearing and varroa control
- distinct honey types can be kept separate
- you can meet consumer demand for fresh, new season's honey.

Not all of a honey comb needs to be capped before it is removed and extracted — frames with at least three-quarters of their area capped are ready for harvesting. But you should be careful not to take frames with fresh nectar that can be shaken out of the cells like water. This dilute nectar has a high water content, and if included with the honey at harvest may increase the water content to the point where the honey may start to ferment later. Some honey types are naturally high in moisture even when fully capped, such as manuka, tawari and kamahi, so be more particular with these types not to include too much nectar or partially processed honey.

It is best to harvest honey on fine, warm days while most of the foraging bees are away from the hive. Normally all honey from supers above the brood nest is harvested. Honey in the brood nest is left for the bees' winter stores; otherwise the colonies will need more feeding in the spring.

Whether you harvest honey progressively or once at the end of the season, the timing of the final harvest is very

Fig. 14.1 Capped frame of honey ready for harvesting.

important now that varroa is present in New Zealand. Most varroa treatments need to be applied when there are no honey supers on the hives.

If you are keeping beehives in areas where tutu bushes are found you should consider harvesting your honey by the end of December. If you are selling honey from such areas you will need to comply with regulations as described in Chapter 20.

PRECAUTIONS WHEN HARVESTING HONEY

You must inspect the brood nest every time honey is taken off a hive, to ensure the colony is free of American foulbrood (AFB). If you harvest honey from diseased hives, any honey contaminated with AFB spores that remains in the combs will spread the disease to the hives that the frames are put back on.

When harvesting honey you should also take precautions to prevent robbing. During the honey flow robbing will not be a problem, as forager bees are fully occupied with gathering nectar. But when the honey flow finishes, the large population of forager bees switches its attention to anything sweet. Once a few bees start robbing, a frenzy soon follows. Colonies close by become aroused and investigate the nearby area, causing a nuisance to neighbours as well as to the beekeeper. Weaker hives and nuclei can be completely stripped of their honey stores by robbing bees.

Be very careful to prevent robbing when you are working hives. Once the honey flow is over, have hives open only for the shortest possible time. Keep harvested honey covered, and take care not to drop honey or scraps of comb onto the ground. Carry a bucket with a lid to contain any comb scrapings or pieces of comb that may have been built in the space between frames.

If robbing gets out of hand in a home apiary, reduce all the hive entrances and, if practical, turn the garden sprinkler onto them. This gentle simulated rain will make most of the bees stay in their own hives.

Also take care not to initiate robbing when you are extracting. Store boxes that contain honey frames, and freshly extracted combs ('wets'), in a bee-proof room. Unless your extraction room is completely bee-tight, do the extracting at night or when it is raining. It is also advisable to kill robber bees rather than let them escape from the extracting room, as they will communicate its location to other bees. Put wet combs back on hives in the late afternoon or evening, or when it is raining.

HONEY HARVESTING METHODS

The object of harvesting honey is to separate the bees from the honey combs and supers. Ways of doing this vary from the very simple to the use of special machinery, and include shaking and brushing, and using bee escapes, fume boards with repellent chemicals, and bee blowers.

Brushing bees

A simple method of harvesting honey from a few hives is to brush the bees off the combs with a special bee brush (available at beekeeping supply firms), though the method does have some drawbacks. If you brush bees off any combs that are not fully capped, or over broken burr comb, the brush quickly becomes clogged with honey and loses efficiency. With brushing, honey is exposed for quite a long period and if this is done after the honey flow, robbing conditions will almost certainly develop. Brushing also makes the bees more aggressive than some other methods.

One way of brushing bees is to use an empty super as a stand for the super being brushed. After checking the hive for American foulbrood, put the empty super down in front of it. Remove a honey super from the hive, place it on the empty super, and quickly cover the hive to prevent robbing.

There is no need to remove frames individually to brush them. If the frames are tightly packed in the box remove one or perhaps two frames, otherwise pull them together on the nearest side of the box. Take the brush in one hand and sweep the bees off the far side of the outside comb, and also sweep those standing on the inside of the box. Push this comb quickly over to the far

Fig. 14.2 Brushing bees off combs over an empty box.

side. With a sweeping motion from end to end of the box, brush the bees off the sides of the two combs that are facing the opening (Fig. 14.2). A quick sweep across the face of each comb will dislodge the bees, and they will fall down inside the empty super and eventually re-enter the hive.

Repeat the process for each comb across the super until all have been brushed free of bees. Put the super on a drip tray on a trailer, truck or wheelbarrow, and cover it with a division board or sack to prevent robbing.

Another method of brushing bees is to remove each honey frame singly, shake most of the bees off, and brush the remainder in front of the hive before putting the frame into another super (Fig. 14.3). Keep exposed frames covered as much as possible.

Brushing combs is fairly slow and is most suitable for a small number of hives, whereas in a large apiary after the honey flow it can provoke severe robbing. Use this method only if you want to harvest a small number of frames, or if you cannot lift a full honey super.

Fig. 14.3 Brushing bees off a comb in front of the hive entrance.

Bee escape boards

A bee escape board is a division board fitted with a non-return passageway for bees to pass through. It is placed underneath any honey supers to be removed, and bees moving down to the brood nest are prevented from returning to the honey supers. In this way, the supers are normally cleared of bees overnight.

When used properly, bee escape boards are efficient and will not cause robbing, provided all the hive parts are bee-tight. They are particularly suitable for urban apiaries. On the other hand, two visits must be made to the apiary, and the honey supers must be lifted twice.

Escape boards can be fitted with corner, Porter or round bee escapes purchased from beekeeping equipment stockists. Use two escapes per board to clear the supers quickly, and in case one becomes blocked. Mark the top side of the escape board clearly to prevent it from being put on upside down.

Cheaper escape boards can be home-made by modifying a division board (Fig. 14.4).

Fig. 14.4 Home-made bee escape.

Drill a 25 mm hole in one corner and make an escape chamber with two pieces of 20 mm x 10 mm timber. These should come together at a tapered point to form an escape tunnel. A simple way of measuring the size of the tunnel is to use an ordinary lead pencil as a gauge. The escape chamber is covered with gauze, or perforated or solid metal. Again, two escapes should be used per board.

An escape board can also serve as an ordinary hive inner cover, provided the escape hole is covered or plugged. If you do this, remove the trap part of Porter escapes and store them when not in use, to prevent bees propolising the springs together.

Before using an escape board check the hive for American foulbrood. When reassembling the hive put the escape board above the brood nest and stack the honey box or boxes on top. If you have very strong colonies and the honey flow is still in progress put a super with empty combs on top of the brood nest, but below the escape board, to accommodate the bees coming down from the honey frames.

When using escape boards, remember:

- The supers being cleared must be absolutely bee-tight, or else the unguarded supers may be robbed out, or at least refilled with robbing bees.
- Bees will not desert any brood or the queen in the supers above the escape board, which is why escape boards work best if you also use queen excluders.
- Bees can also be reluctant to clear uncapped honey combs. A bee brush will be needed for the last remaining bees.
- Escapes may not work as well when nights are very hot and humid.

Fume boards

Fume boards resemble hive lids but are lined with absorbent material such as towelling or newspaper, which is sprinkled with a repellent chemical (Fig. 14.5). The top of the board is painted black to absorb the sun's rays. The absorbent material can be held in place with wires stretched from corner to corner.

Supers of honey can be cleared of bees by taking off the hive lid and inner cover,

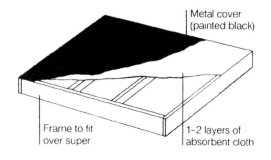

Fig. 14.5 Construction of a fume board.

and replacing them with a fume board. After a few minutes the bees will move down into the brood nest. The advantage of the fume-board method is that only one trip to the apiary is necessary.

Benzaldehyde or artificial oil of almonds was used in the past by many beekeepers and may still be available, although it is not registered for use in beehives. Benzaldehyde has a pleasant smell but is very corrosive to plastic and will burn exposed skin, so always use rubber gloves when handling it. It is also reasonably effective in cold weather. Bee-Quick® is a registered bee repellent product that contains a non-toxic blend of oils and herbal extracts. Bee supply stockists should be able to advise on availability of repellent chemicals.

To use a fume board, remove the lid and inner cover from a hive. Smoke the bees in the top super fairly heavily to start them moving downwards. Apply the amount of chemical recommended on the label directions, put the fume board on the hive and wait for a few minutes. A quick check will show when the top super is clear of bees, at which point it should be removed. Place the fume board on the next super if several are to be cleared, and repeat until you reach the brood boxes.

When harvesting honey from an apiary of more than about 15 hives, use about three or four fume boards so honey can be removed more or less continuously. 'Leapfrog' the boards down the row of hives as supers are cleared, so each board has several minutes to work on a super.

Fume boards are efficient tools for harvesting honey, but using them requires skill. Common problems include not smoking the bees heavily enough before the board is put on, or using too much repellent, as the bees can become stupefied and will not move. Experiment and persevere until you find the right combination of chemical, smoke and time for maximum effectiveness. A bee brush will sometimes be needed to move any bees that have not left the honey supers.

Wash fume boards, especially the absorbent material, thoroughly with water before storing them.

Bee blowers

Beekeepers with larger numbers of hives usually use a bee blower to force bees out of honey supers. These are fast and efficient and work in all weathers. Both petrol-driven and electric bee blowers are used, and specialised petrol-driven blowers can be bought from beekeeping-supply firms. Two-stroke leaf blowers are effective but can be noisy and unpleasant to use for long periods.

The noise and vibration associated with any type of motorised bee blower has led many beekeepers to change to hand-held electric industrial blowers. These are easy and quiet to use and are switched on only when needed, instead of running continuously. The electricity can be supplied by a power converter running off batteries, or by a petrol-driven generator.

Bee blowers should be used in conjunction with a stand to put the supers on (Fig. 14.6). Place the stand in front of the hive so that the bees will be blown onto the ground outside the entrance, from where they soon re-enter the hive. Never place boxes on their ends to blow bees out into the air, as the ground in the apiary will soon become covered with a dense carpet of confused bees.

Manley frames, with their wide spacings, are easy to blow bees from. Hoffman flames should be pushed to one side of the super and worked across as each space is cleared.

Fig. 14.6 Using a bee blower.

Honey composition and properties 15

Honey has a long heritage as a prized natural food, and harvesting this wonderful product is the reason that many people take up beekeeping. However, in order to process honey without damaging its flavour or changing other characteristics it is important that you know something about its composition and properties.

WHAT IS HONEY?

The definition of honey set down in the international food standard Codex Alimentarius reads: 'Honey is the natural sweet substance produced by honey bees from the nectar of plants or from secretions of living parts of plants or excretions of plant-sucking insects on the living parts of plants, which the bees collect, transform by combining with specific substances of their own, deposit, dehydrate, store and leave in the honey comb to ripen and mature.'

SOURCES OF HONEY

The most common raw material for honey is nectar obtained from the blossoms of flowering plants (Fig. 15.1). This is a watery solution of various sugars, mainly sucrose, produced by organs called nectaries inside flowers. Honey bees gather nectar from a wide range of flowers, and in New Zealand surplus honey comes from pasture plants as well as trees and shrubs (see Chapter 6).

The other main type of raw material is honeydew, a sugary product of sap-sucking insects living on plants. This honeydew is collected by bees, modified and ripened, and stored as honey. The

Fig. 15.1 Manuka flower.

Fig. 15.2 'Whiskers' or tubes from honeydew scale insects in beech forest.

Fig. 15.3 Foraging bee collecting honeydew from a beech tree.

most abundant source of honeydew in New Zealand is scale insects living inside the bark of native beech trees (*Nothofagus spp.*), particularly in the northern half of the South Island. The honeydew is passed out through fine tubes attached to each scale insect (Fig. 15.2), and bees collect it either from the tubes or from where it falls (Fig. 15.3). Beech honeydew is the basis for significant beekeeping operations, producing a valuable honey.

Honeydew is also produced by different insects on trees such as lime, oak and sycamore, and plants like blackberry, but in New Zealand these sources are not significant enough for bees to produce surplus honey from regularly. Bees will occasionally collect sap secretions from cut maize plants.

Toxic honeydew is produced when the vine hopper *Scolypopa australis* feeds on the toxic native plant tutu (*Coriaria arborea*). Honey bees may collect this honeydew when there are limited or no nectar flows. Significant amounts of toxic tutu honeydew honey can be stored by bees in the eastern Bay of Plenty, the Coromandel Peninsula, East Cape, Poverty Bay and the Marlborough Sounds. For information about regulating the harvesting and sale of at-risk honey, see Chapter 20.

Minor sources of nectar include nectaries on parts of plants other than flowers (extra-floral nectaries). Occasionally bees may also be seen collecting juice from broken, over-ripe fruit. However, it is believed that bees do not make the initial opening, which is usually done by wasps, birds, or by the fruit itself bursting.

COMPOSITION

The major components of nectar or honeydew are water and sucrose, which is the same chemical that gives sweetness to the common white table sugar made from sugar cane or sugar beet. When bees convert nectar or honeydew to honey they reduce the water content, and add the enzyme invertase to convert most of the sucrose to glucose and fructose. This process is called 'ripening' the honey. Other chemical changes take place during storage in the hive, processing by the beekeeper, and subsequent storage of the extracted product.

The average chemical composition of floral honey is given below in Table 15.1, but honeys from different floral sources vary significantly in chemical composition and physical properties.

Table 15.1   Average composition of floral honey

	%
Glucose	31.3
Fructose	38.2
Maltose (and other reducing disaccharides)	7.3
Sucrose	1.3
Higher sugars	1.5
Total sugars	79.6
Water	17.2
Organic acids	0.6
Minerals	0.2
Other components	2.4
Total other	20.4

PHYSICAL PROPERTIES

Some of the titles listed at the end of this book cover the chemistry and physics of honey in detail, but here we outline only some characteristics of honey that you need to keep in mind when extracting and processing your crop.

Viscosity

This is honey's resistance to flow. It is what beekeepers mean when they talk about its 'body', and is probably the most important factor affecting honey processing. Viscosity depends mainly on temperature, but is also influenced by floral source and water content. (A different property related to the 'body' of honey, thixotropy, is discussed below.)

Honey is much easier to handle if it is heated from room temperature (15–20 °C) to about 30 °C. Further reductions in viscosity between 30 °C and 45 °C are small, and are only important for operations such as the pumping, straining and filtering of bulk honey.

Thixotropy

Thixotropic honeys exist in a jelly-like state, which can be changed to a liquid state by physical agitation. If left undisturbed the liquid honey will revert to the jelly state. Thixotropy is due to the presence of a colloidal protein in the honey. Honeys from the native plants manuka (*Leptospermum scoparium*) and kanuka (*Kunzea ericoides*) are thixotropic, as is honey from ling heather (*Calluna vulgaris*), an introduced plant found in parts of the central volcanic plateau of the North Island.

Thixotropic honeys are hard to extract. If you encounter thixotropic honey you will have to use the technique described in the next chapter, produce comb honey, or accept that a portion or your crop will not be extracted.

Water content

Honey is one of the most stable foods known, and if stored correctly can be eaten decades after being harvested. Proper storage is essential, and preventing honey from absorbing excess moisture is critical if it is to remain in good condition.

Honey usually absorbs water from its surroundings, so the water content tends to increase with time unless honey is stored in airtight containers. Water absorption can put honey at risk of fermentation, and if the water content exceeds about 19 per cent, the sugar-tolerant yeasts present in honey will initiate fermentation unless the honey is pasteurised to kill the yeasts, kept in cool storage or frozen. Honey with less than about 17 per cent water content rarely ferments, despite the yeasts still being present.

Refractive index

The refractive index denotes the extent to which honey bends a ray of light, and provides a simple, fast and very accurate measurement of the water content of honey since the two properties are directly related. Hand-held refractometers specially calibrated for honey are very useful for

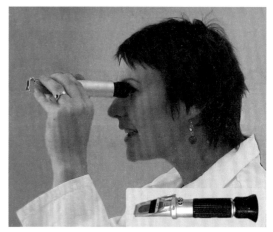

Fig. 15.4 Using a refractometer.

checking on the water content (Fig. 15.4). Brix refractometers used to measure the sugar content of fruit are not suitable for this task.

Colour

Honeys from different plants vary greatly in colour, ranging from virtually colourless (white) through shades of yellow, amber and brown to almost black. Occasionally honeys with red or green tinges are produced. Honey darkens with heating and prolonged storage. Table 15.2 shows the colour categories of some important local honeys. The book *Nectar and pollen sources of New Zealand*, which is among those listed at the end of this book, discusses the colour of honey (and pollen) for almost every floral source in New Zealand.

Table 15.2 Colour of some New Zealand honeys

Colour	Floral source
White	white clover, rata, pohutukawa, southern rata, canola, tree lucerne, red clover, blackberry, lucerne, viper's bugloss, thistles
Extra light amber	willow, five finger, pip and stone fruit, tawari, most yellow-flowering weeds such as catsear and hawks-beard, native fuchsia (kotukutuku)
Light amber	dandelion, barberry, kowhai, some manuka, kanuka, heath, buttercup, koromiko, *Lotus major*, kamahi
Medium to dark amber	rewarewa, some manuka, heather, gums, pennyroyal, beech honeydew

Colour is important for marketing purposes, and an instrument known as a Pfund grader is often used to measure it. This device is calibrated in an arbitrary scale of millimetres, measuring the distance a sample of honey has to be moved to match the colour of a wedge of coloured glass.

Acidity

There are small quantities of organic acids present in honey, principally gluconic, lactic and pyroglutamic acids, that make honey quite acidic. Honey has a pH of 3.6 to 4.2, which is nearly as acidic as vinegar. Honey's high sugar content, however, ensures that it tastes sweet, not sour.

Honey's high acidity means that retail packs have to be made of glass or food-grade plastic, while bulk steel drums have protective coatings on the inside. Honey will corrode unprotected steel and over time will pit concrete floors.

Thermal conductivity

Honey is a very poor conductor of heat, and must be heated carefully to avoid burning it. This is particularly important when processing thixotropic honeys, as applying excess heat will not necessarily make the honey more manageable.

Granulation

There is naturally more glucose in most honeys than can stay dissolved, so the glucose will crystallise out of solution in a process called granulation.

The rate of granulation varies with the floral source of the honey, and depends on its glucose content and moisture level. Some New Zealand honeys granulate rapidly, for instance rata, brassicas (especially canola), and pohutukawa. Unless these are extracted within a few weeks of being capped the honey may granulate in the comb. Other honeys may take much longer to granulate (e.g. viper's bugloss or beech honeydew). Thixotropic honeys such as manuka, kanuka and ling heather still granulate, but with very large crystals. The temperature at which honey is stored also greatly affects the speed of granulation.

Granulation is a natural phenomenon in most New Zealand honeys. The slower the granulation, the larger and coarser the resulting glucose crystals become. Coarse crystals are unappealing to most consumers and can be avoided by producing liquid or finely granulated honey, which is explained in the next chapter.

Calorific value

Honey is a concentrated source of energy, since about 96 per cent of its solids are sugars. Its energy value is about 1270 kJ (300 calories) per 100 g, or 250 kJ (60 calories) a tablespoon.

Honey and diabetics

Diabetics especially should be wary of claims that honey is a 'sugar-free' food. It is not, because honey is composed almost entirely of sugars and water, including a high proportion of glucose. Honey has about the same sweetening power as cane sugar (sucrose) on a weight-for-weight basis; however, it contains only about 31 per cent glucose compared with the 52 per cent cane sugar yields after digestion in the intestine. Honey gives the same sweetening power at a lower glucose 'price' than cane sugar, but any substitution of honey for sugar should only be done on medical advice.

Antibacterial activity

Honey has been used as a medicine since ancient times, and the nature of its antibacterial properties has been under investigation for more than a century. Research at the University of Waikato in New Zealand since the early 1980s has revealed significant new information about the way honey's antibacterial properties vary with its floral source, area of production and subsequent storage and heating (see www.waikato.ac.nz and in particular bio.waikato.ac.nz/honey).

There are several reasons for honey's activity against microbial organisms. As a saturated or supersaturated solution of sugars, honey has a high osmotic (water-withdrawing) effect that restricts the activity of many bacteria. Because honey is quite acidic it inhibits many disease-causing organisms.

In honeys other than some manuka honeys, the major antibacterial activity has been found to be a result of hydrogen peroxide produced enzymically in the honey. The enzyme glucose oxidase is secreted from the worker bee's hypopharyngeal gland into the nectar during the honey-ripening process, and the hydrogen peroxide produced helps to preserve the honey during ripening. Fully-ripe honey has a negligible level of hydrogen peroxide, but on dilution the activity increases hugely to a level that is antibacterial but which does not damage tissues. Honeydew honey has particularly high antibiotic activity due to hydrogen peroxide.

There have also been reports of antibacterial substances in honey other than hydrogen peroxide, though often these are present at levels too low to account for significant activity. But some honeys from the native New Zealand plant manuka (*Leptospermum scoparium*) have been found to have a high antibacterial activity, much of it 'non-peroxide' activity. Manuka honey that has been laboratory tested for antimicrobial activity and properly stored is used as an antibacterial agent, especially in treating stomach ulcers and diarrhoea, and dressing burns and other wounds. A wound dressing using manuka honey has been approved by the US Food and Drug Administration, and is used in hospitals in many countries.

Manuka honeys with a high antibacterial effect are said to have a 'unique manuka factor', or UMF, which is expressed by a series of numbers after the UMF trademark, e.g. UMF 5+, UMF 10+. The numbers relate to the antimicrobial activity shown by laboratory testing, and are related to an equivalent solution of the antiseptic phenol (carbolic acid). For example, a honey with a UMF of 10 would have the same ability to inhibit bacterial growth as a 10% solution of phenol. The initialisation 'UMF' is now a trademark of the Active Manuka Honey Association. Not all manuka honey shows this antibacterial activity.

Further research in Germany and at the University of Waikato in New Zealand has shown that a chemical called methylglyoxal can build up in manuka honeys over time, and this contributes to some of the observed antibiotic effects. University of Waikato researchers have also identified a chemical called dihydroxyacetone (DHA) found in some fresh manuka honeys. As the honey ripens, the DHA converts to methylglyoxal, and the antibacterial activity of the honey increases over time. Some beekeepers believe that heating the honey hastens this process, but researchers have found that this is not the case. In fact, prolonged heating can increase the levels of hydroxymethylfurfural or HMF, which may make the honey suitable only for industrial uses if levels exceed those specified in the Codex Alimentarius standard, or set by importing countries (especially in the EU) or importers.

University of Waikato researchers also identified DHA in the nectar of some manuka flowers, and this can be used to identify plant specimens for possible propagation for manuka honey production. Testing manuka honey for DHA can also tell beekeepers if their honey is likely to develop antibacterial levels or not. This will help in deciding if the honey should be sold as table honey or reserved for medicinal purposes.

Some other New Zealand honeys such as rewarewa have been shown to be high in antioxidants, and these may also add to the antibiotic profile. New research is focusing on the anti-inflammatory properties of some honeys.

Fig. 15.5 Manuka honey — now a high-value product.

Extracting and processing honey 〈16〉

Extracting honey is the process of removing it from the comb. This is most commonly done with a centrifugal honey extractor, although there are other methods. This chapter also deals with processing the honey after extraction to yield a good-quality product.

Beekeepers with only a few hives usually extract their honey in a shed, garage or kitchen. Provided this is only for your own consumption there are few restrictions on your activity. However, if you intend to sell or barter your honey (or other bee products) for human consumption, you will have to comply with a range of requirements relating to things such as safety, wholesomeness and labelling. These are covered in detail in Chapter 20.

Home honey extraction should be done in a place that is bee-tight, or carried out after dark when robber bees are not flying. The room should be free of dust and have easily-washable walls and floor, a convenient water supply and electricity.

If you plan to extract honey in the kitchen, first reach a clear understanding with other users about cleaning up afterwards. Layers of newspaper covering the floor will catch some honey, but not the fine droplets flung out of the extractor onto the walls. In addition, propolis will inevitably get stuck on surfaces — it can be very difficult to remove, and it readily stains.

An extractor is necessary for efficient processing of honey, but the cost of a new two-frame model — the smallest type available — is at least twice the price of a complete hive. If you have only a few hives it is best to borrow an extractor, hire one from a local beekeeping club, or buy shares in one with other beekeepers. Some hobbyist beekeepers have made their own extractors, and this is an alternative for someone with practical skills.

WITHOUT AN EXTRACTOR

If you only have one hive you can extract honey using an old spoon sharpened on one edge to scrape the honey and wax from each side of the foundation sheet, or midrib, of the comb. Plastic frames are ideal for this as it is impossible to break through the midrib.

Collect the resulting mass of honey and wax, and strain or squeeze it through a piece of fine plastic mesh, muslin or cheesecloth that has been lightly moistened. This method is slow and messy, and leaves both a lot of honey behind in the wax cappings and a lot of wax in the honey. Use it only if an extractor is not available.

Section or cut comb honey does not need extracting, but producing this is best done in conjunction with extracted honey rather than as a complete alternative. Comb honey is discussed in the next chapter.

WITH AN EXTRACTOR

Temperature

Honey is very sticky or viscous at low temperatures, and should be kept between 30 and 40 °C to make handling easier. Temperatures above 40 °C give little further reduction in viscosity and are liable to burn honey if not applied carefully. Fig. 16.1 shows the effect of moderate heating on honey viscosity.

The easiest way of controlling the temperature of a few boxes of honey is to extract them immediately after harvesting from the hives. This is particularly effective if the extracting room is kept warm. If honey must be kept overnight before extracting, keep the room where it is stored at around 30 °C with a thermostatically controlled heater.

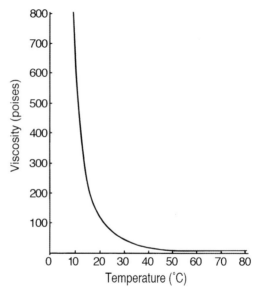

Fig. 16.1 The effect of moderate heating on the viscosity of honey.

Various devices can be made to heat small stacks of honey supers. These usually consist of an empty super or supers fitted with a fan heater or low wattage incandescent light bulb, on which several full boxes of honey are stacked. Hot air passes up through the honey and keeps it warm. The danger in using such devices is that beeswax combs start to sag at just over 50 °C and melt completely at 64 °C. Even if the heater is fitted with a thermostat that does not fail, localised hot spots can develop, resulting in a flood of honey and molten wax and perhaps a fire.

Uncapping

The first step in extracting honey properly — with as little damage as possible to the combs — is to remove the wax cappings over the honey cells. This is done with a hot, sharp knife.

The most basic uncapping setup is to use a long kitchen knife kept hot by standing it in simmering water, and having a pair of knives ensures a hot blade is always available. Knives with serrated edges, such as bread knives, work best. Always wipe the knife dry before you use it each time to avoid adding water to the honey. Bee equipment stockists also sell plain 250 mm uncapping knives, which can be used in the same way.

Fig. 16.2 Processing equipment in a small-scale honey house.

A more convenient way of uncapping is to use a steam or electrically heated uncapping knife, available from a bee-equipment supplier. Electric knives are fitted with a thermostatically controlled element, which can be adjusted with the small grub screw on the back of the blade. They should be operated so that honey left on the blade just sizzles after a few seconds. Remember that an electric uncapping knife is not particularly robust, and should be handled gently.

Steam-heated knives are the same size as electric knives, but instead of a heating element inside have a copper tube soldered to the back of the blade. They need to be connected to a simple steam generator made from something like a pressure cooker, electric kettle or wall-mounted water heater. The steam generator must be fitted with a safety valve to prevent accidents.

Fig. 16.3 Uncapping honey combs with an electric uncapping knife.

Uncapping is easiest if eight frames are used in each box, so you can uncap by cutting right down to the level of the top and bottom bars (Fig. 16.3). Although this speeds up the process, as much as one-third of the honey crop can be bound up with the cappings. In this case an efficient processing system is needed to handle the cappings quickly and without damaging the honey. If your system is fairly basic, uncap the combs to as shallow a level as possible to reduce the amount of honey that later has to be separated from the cappings.

When uncapping with a hand knife, support the end of the frame on the point of a nail driven through a piece of wood. This makes the frame easier to hold at the correct angle, and to turn when you want to uncap the other side. Put this support across the top of a basket into which the cappings can fall and drain. A simple cappings basket can be made by nailing a queen excluder to the bottom of an empty super. Fine plastic mesh gauze can be placed on the excluder if desired.

Use the knife with a gentle sawing motion, either downwards or upwards depending on your preference. If slicing up be careful to keep your hand holding the comb away from the blade, and if cutting down try not to hit the support bar too often, especially with an electric knife. Depressions in the comb can be uncapped with the tip of the knife or scratched with a fork or cappings scratcher, so that all the honey will later come out in the extractor.

Cappings scratchers are available from bee equipment stockists and can be used to scrape entire frames as well as portions missed by uncapping knives. They work reasonably well but damage the comb, and add a lot of small pieces of wax to the honey that need to be strained out.

Uncapping knives can be mounted horizontally with one blade uppermost, so that the comb is moved rather than the knife. Some beekeepers find this setup less tiring than using a hand-held knife for long periods, especially if the mounted knife is vibrated back and forth by an electric motor.

Commercial beekeepers use uncapping machines with chain feeds to move the frames past oscillating, heated knives. Some new machines made in New Zealand especially for manuka uncap down to the midrib of the comb. These machines uncap and extract in the same motion, and the large volume of wax and honey produced is then processed in a spinning separator or continuous flow press.

Place uncapped combs on a stand, such as a plastic basin or clean honey super fitted with a drip tray, until they are put into the extractor. Handling cappings once they are removed from the combs is dealt with in Chapter 18.

Extractors

Extractors vary in size, from small hand-driven models holding two frames to large electrically driven types accommodating nearly 200 frames. A small domestic honey house will probably be fitted with a two- or four-frame tangential extractor, so called because the frames sit at a tangent to the direction of rotation. Small extractors of this size are usually hand-driven (Fig. 16.4), although the job is made considerably easier if they are powered by an electric motor.

In any sort of tangential extractor only one side of the comb is extracted at a time. When the first side of the comb is extracted there is a rush of honey to start with as the bulk of it spins out, but the flow lessens as the remainder is extracted. With the weight of the full side pressing on the empty side of the comb, breakages often occur at this point. This happens especially in new combs with wax foundation, whereas plastic foundation is much more robust.

To avoid breaking wax foundation combs, spin the first side until the initial rush of honey has come out. Stop the extractor and reverse the combs — in a simple extractor by first removing them, but in larger extractors by using the reversible baskets. Spin the second side to remove all the honey, reverse the combs again and complete the first side. This procedure also helps to prevent the extractor wobbling.

In a commercial honey house nothing smaller than an eight-frame extractor is suitable to achieve satisfactory throughput. Where the common honey types are readily extractable, semi-radial or even radial extractors are used. In radial extractors the combs sit along a radius of the circle, so both sides are extracted at once (Fig. 16.5). Semi-radial extractors have baskets and need to be reversed to extract both sides. Radial extractors are satisfactory only for use with light-bodied honeys. They extract more combs at a time than tangential extractors, but take longer to spin out all the honey. Tangential and semi-radial extractors are best for heavy-bodied honeys.

Fig. 16.4 Hand-driven, two-frame tangential hobbyist extractor.

Most extractors of all three types have a vertical shaft and are loaded and unloaded from the top. There are also horizontal extractors that are loaded from the side, but because of their size they are best suited to large-scale operations.

Thixotropic honeys can be extracted more suc-cessfully if they are agitated just before extraction with a honey pricker. Simple prickers can be bought from beekeeping supply firms — they look like paint rollers studded with nails and are rolled

Fig. 16.5 Motorised four-frame radial honey extractor.

several times up and down both sides of the comb. The spikes agitate the contents of each cell and liquefy the honey. Commercial prickers have banks of spring-loaded blunt plastic needles that enter and exit the cells several times per cycle.

Removing debris

Honey running from the extractor will contain air bubbles, wax and other debris. If granu-lation does not occur rapidly, the debris can be removed by leaving the honey in a tank for 48 hours and letting the material rise to the surface. The tank should not be emptied completely

Fig. 16.6 Heated baffle tank for removing impurities from extracted honey.

until the end of the extraction so that the surface scum is not mixed with the honey. Tanks should always be covered to keep out flying insects and also to reduce the uptake of moisture by the honey.

It is best to supplement this process by first running the honey from the extractor into a sump, which is usually fitted with baffles (Fig. 16.6). The wax and other floating debris rises to the surface only if the honey is warm, so the baffle tank must be heated by a water jacket, heating cables or a heater placed underneath.

Often the honey is further strained as it goes into the tank. This is achieved by running it through a stocking or nylon mesh bag, which should be suspended in the honey in the tank to prevent premature clogging. More sophisticated commercial operations use stainless steel in-line strainers, rotating stainless steel cylindrical strainers, plastic irrigation filters or centrifugal separators.

Air bubbles in honey make it look less attractive and increase the risk of fermentation. Many beekeepers take pains to remove air bubbles from honey in a baffle tank, only to reintroduce them by allowing the honey to drop from a height into a storage tank. Honey should always run down a chute or the side of the tank so few air bubbles are incorporated.

Honey that is to be sampled and tested for the presence of antibiotic properties or toxins from tutu should be thoroughly mixed in a tank for many hours before sampling, to ensure the sample is homogeneous.

Tanks

Tanks can be made of any material that will not taint the honey. Stainless steel, polypropylene, fibreglass, or any material that is lined with a food-grade epoxy reaction lacquer is suitable. Honey is quite acidic and reacts with unprotected steel to form iron tannate, a bitter-tasting black substance. If you intend to sell your honey to a packer, ask for advice on suitable containers as there may be restrictions for export. Remember to keep tanks covered

Fig. 16.7 Small-scale honey packing machine.

to prevent bees and wasps becoming trapped in the honey, and also to prevent moisture absorption.

PROCESSING HONEY

When processing honey, take into account its physical properties (see Chapter 15), especially its viscosity and heat sensitivity. Also remember that honey is usually a super-saturated glucose solution — that is, it contains more glucose than can stay dissolved. This means that the glucose will come out of solution, and as a result most New Zealand honeys will crystallise or granulate over time at room temperature.

Liquid honey

The simplest and surest way to produce liquid honey is to pack the honey into small containers and store them in a domestic freezer. At this temperature granulation is greatly retarded, and when the honey is returned to room temperature it will still be liquid. This method causes the least damage to honey. If you are going to label the containers it is best to do this after the freezing process.

There is another method, which is more complicated for hobbyist beekeepers and more likely to spoil the honey. You can pack honey into heat-stable plastic buckets or glass jars, allow it to granulate naturally, and melt it out when liquid honey is required. This must be done very carefully, as it is easy to damage the honey.

You can melt out the containers of honey by immersing them in a water bath fitted with a thermostatically controlled heater set at 60 °C, or by heating them in an old refrigerator fitted with a low-power incandescent light bulb (start with 40 W) and a thermostat. Do not use direct heat or the honey's taste and colour may be affected. Because honey is a poor conductor of heat it can burn easily, so it should be stirred at intervals during the melting process.

If honey does granulate in glass jars it can be melted in a microwave, but again this is a risky business. Remove the lid and try small bursts of power for 15–30 seconds. Honey heats very quickly so keep testing and stirring as the crystals melt. If there has been some fermentation in the honey, very full jars may overflow. Plastic jars will melt.

Honey sold commercially as liquid honey has usually been selected from sources that are slow to granulate, and the honey is further treated to delay the formation of glucose crystals. First, particles around which a glucose crystal would grow are removed by careful filtration or fine straining. Second, the moisture content is adjusted to between 18 and 19.5 per cent, usually by blending honeys. Finally, the honey is pasteurised or briefly flash-heated to around 70 °C to destroy yeasts that might cause fermentation, and to dissolve any tiny glucose crystals that may be present. It is then cooled immediately to around 30 °C and packed in dust-free jars. Most packed liquid honey has a shelf life of only three to 12 months before granulation starts to occur, though this depends a lot on the honey's floral source and the temperature at which it is stored.

Granulated honey

To avoid slow granulation and large sugar crystals, most retail honey in New Zealand is granulated to a smooth, fine-grained state before packing. This is termed granulated, crystallised or creamed honey. No pasteurising is required, and no additives are used. Although granulated honey may have different physical properties from liquid honey, it is chemically and nutritionally the same.

The technique for producing fine-grained honey is sometimes called the 'Dyce process' overseas, after the Canadian who described it in 1931. However, the technique was well known in New Zealand before that date, and had been described by Isaac Hopkins in articles published as early as the 1880s.

The granulation method involves adding fine-grained starter or seed honey to liquid honey. The many small glucose crystals in the starter act as nuclei on which glucose crystals will grow in the liquid honey. The mixture is stirred thoroughly to ensure that the starter is evenly dispersed, then held at the temperature at which crystal growth is most rapid.

To granulate honey

- Select a very fine-grained honey as a starter, as the final product can only be as smooth as the starter used. Its colour and flavour should be similar to, or lighter than, the bulk honey, as a strong-flavoured starter will dominate the flavour of a mild honey. Starter honey can be some of your previously granulated honey, or crystallised honey purchased from a shop.

- Soften the starter by placing it in the hot-water cupboard or another warm place for 24 hours. Never use direct heat, as the glucose crystals could be melted and the starter made useless.

- Add five to 10 per cent by weight of the softened starter to the bulk honey. Be sure to mix it in very thoroughly, as it is easy to give up before the job is completed and incomplete mixing will result in very uneven granulation. Complete mixing may be easier if the starter is added to some liquid honey to make a 'paste' before being mixed with the remainder of the liquid honey.

- Stirring the starter in is easiest if the bulk honey is warm (around 20 °C), but it must not be much hotter or it will dissolve the glucose crystals in the starter. A large industrial drill with variable speed control and a paint stirrer attachment can make this job easier.

- Pack the resulting mixture into containers.

- Store the containers at 12–14 °C, the temperature at which crystal growth is most rapid and a very fine texture is achieved. A cool, dry basement may be used for this, but in summer it is difficult to find places with the correct temperature. Although a domestic refrigerator is a little too cold, it gives better results than a warm room during summer.

- Commercial honey packers leave the bulk granulating honey for two or three days at 12–14 °C before packing, as pumping or slow mechanical stirring during this time results in a smoother product. This is not possible without cooled tanks or a cool room.

- About a week after packing, the honey will have finished granulating. It can then be stored in any dry place, ideally at around 17 °C. If left longer than a week at 12–14 °C, the honey shrinks away from the sides of the jars and glucose crystals grow in the air spaces, leading to a condition called 'frosting'. This is not harmful but it does not look very appealing.

- If the honey is too hard when it is used, it may be softened in a warm place (e.g. a hot-water cupboard) for 24 hours first. But once granulated honey has been softened, it will not go hard again unless it is melted, new starter added and the process described above is repeated.

- You can use a two-step granulation process to bulk up a small amount of starter to the required 5% of the bulk honey to be granulated. Buy a small amount of granulated honey and mix it with enough liquid honey to make about 5% of the total bulk honey. Granulate this starter mix at 10–14 °C until it has set, and then use the granulated starter with the larger bulk of liquid honey in the normal way.

- Granulated honey usually looks lighter in colour than the liquid honey it came from. This is due to air pockets and the glucose crystals reflecting light, just as ice looks white. This is probably why the urban myth still persists that icing sugar has been added to crystallised honey. Just as ice can be melted and refrozen, so granulated honey can be melted, seeded with crystallised honey and granulated again, proving that nothing but honey has been added.

MARKETING HONEY

Regardless of how you intend to dispose of most of the honey crop, some will always be needed for household use, gifts for people whose land is used for apiary sites, and similar purposes. This honey can be packed in any clean plastic or glass containers with firmly fitting lids.

Selling honey on a 'fill your own container' basis is a convenient method of sale. Local sales to shops, wholesalers, or through offices and clubs are also suitable ways of disposing of part or all of a honey crop, but remember that the legal standards set out in Chapter 20 must be met. If you intend selling honey to shops or produce stalls, present it in an interesting way and be sure that the consumer is able to rely on a consistent, high-quality product (Fig. 16.8).

Bulk honey is usually sold to a private packer, exporter or honey cooperative in 200-litre honey drums, or less commonly in plastic Pallecons. Handling these containers without special equipment is impossible, as drums weigh up to 330 kg and Pallecons well over one tonne. Twenty-litre plastic pails are more suitable for a small business. Part-time beekeepers may find it worthwhile to pool their crops to make up a larger order, or find a beekeeper who is prepared to extract honey under contract and purchase the honey as well.

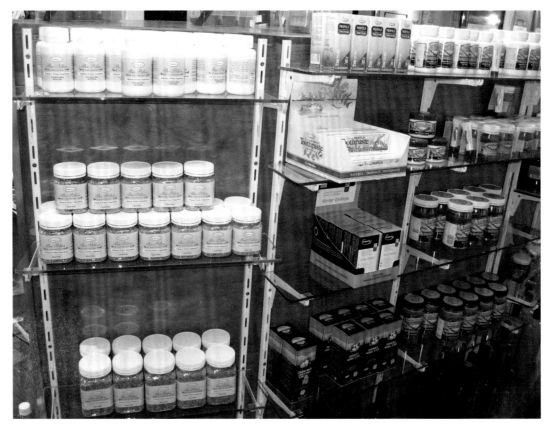

Fig. 16.8 Attractive display of bee products in a specialised honey shop.

Comb honey ⟨17⟩

Comb honey is just as the bees made it, and is preferred by some beekeepers because it retains all the subtle flavours and aromas of honey straight from the hive. Comb honey can be a convenient product for beekeepers without an extractor, or for those who produce thixotropic honeys such as manuka or ling heather. Remember, though, that comb honey should not be produced for human consumption if there is a risk of the bees collecting toxic honeydew from the native plant tutu (see Chapter 20).

The term 'comb honey' includes three different products:

- cut comb honey — small squares or rounds of honey that are cut out of ordinary frames by the beekeeper and packed in plastic containers

- section honey — the combs made by bees in special wooden or plastic holders

- chunk comb honey — pieces of honey comb placed in a jar that is filled up with liquid honey.

Comb honey is more expensive per kilogram than extracted honey. This is because when producing comb honey a beekeeper's yield per hive is lower, hive management is more intensive, and packaging costs are higher.

COMB HONEY EQUIPMENT
Cut comb honey
Cut comb honey is probably the easiest form to produce and package, as no special frames are needed and the plastic containers are readily available. Cut comb can be produced in any size of frame, but the best results are achieved with full-depth frames fitted with a 15 mm-deep top bar. With ordinary full-depth Hoffman frames, use nine or 10 frames per

box, spaced evenly. Some commercial beekeepers specialising in comb honey production use a special cut comb frame, which is self-spacing at 10 to the box and is designed to fit in a 150 mm-deep cut comb super.

Use thin-super foundation in frames for cut comb honey, with four wires per full-depth frame. This grade of foundation is very brittle when cold, so leave it at room temperature (around 20 °C) for 24 hours before handling. It helps to fasten the foundation to the top bar by pouring molten wax into the groove. Foundation should not be put into the frames too soon before the honey flow.

A maximum of four squares can be cut from a cut comb frame, and a maximum of eight from a full-depth frame. Round pieces of cut comb are also an option.

Section honey

Sections can be produced in half-depth or full-depth supers. Management is easier and a better product obtained with half-depth boxes, but these have little use other than for sections. Full-depth supers are more versatile. Each section box is filled with seven section frames or holders. A half-depth section frame holds four wooden sections measuring 108 mm square, and a full-depth frame holds eight (Fig. 17.1).

The traditional wooden sections that you assemble yourself are still available. The sections with a split top bar (Fig. 17.2) are easiest when producing small numbers. Before assembling them, lay a quantity of unfolded sections on their side and pour boiling water on the V-shaped grooves. Take care not to wet the rest of the sections, especially the dovetailed ends. After a few minutes, when the wood has softened, start folding the sections into squares. Put only one half of the top piece in place and hammer it home — leave the other half until later.

Fig. 17.1 Section honey ready for harvesting.

Standard-size, half-depth thin super foundation is used for wooden sections, even when these are for full-depth boxes. It should be cut into pieces 95 mm wide, using a mitre box like that shown in Fig.17.3. The sides of the mitre box are about 60 mm deep, and have four saw cuts to guide a cutting knife.

Put about 10 strips of foundation into the box and cut them with a sharp knife.

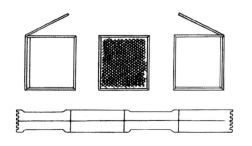

Fig. 17.2 Wooden section.

A straight-edged knife will cut best if it is dipped frequently into soapy water.

Fit the cut pieces of foundation into the slot in the top of the section, and hammer home the other half of the top piece. Remember to put the foundation in with the rows of cells running horizontally so that each has an apex, or pointed end, at the top and bottom. Do not put foundation into frames earlier than a month or so before the honey flow, as it can warp with temperature changes.

Fig. 17.3 Cutting foundation in a mitre box.

Fit the sections into the frames with the scalloped sides top and bottom, putting the end sections in before the two middle ones. Each frame has a tin or hardboard separator, which stops the bees attaching sections together and helps them to build level faces to the sections. All the separators must face the same way so that none touches another back-to-back.

Section frames with hardboard separators are a tight fit at seven per super, but those with tin separators may be loose-fitting. If using the latter, push all the frames together, possibly using wooden wedges in the outside gaps to keep them in this position.

COLONY MANAGEMENT FOR PRODUCING COMB HONEY

Some beekeeping books contain long and complicated instructions for manipulating hives for comb honey production. Many of these methods no doubt work, but there are easier ways of producing a high-quality product.

The best comb honey, either sections or cut comb, is produced by a very strong colony that fills the brood boxes and has already started gathering surplus honey. This means that the hive has a first extracting super on that is partly full. If a comb honey box is put on before the main honey flow starts, the bees will chew holes in the foundation, or it may warp.

When the main honey flow starts in earnest, in late December to early January in most areas, take the extracting super off and replace it with a comb honey super. The partly-filled extracting box can go onto another hive, after you have checked the brood nest for American foulbrood and also shaken most of the bees out. Alternatively it can stay on the hive above the comb honey super, unless it contains dark combs, as the bees might incorporate some of this dirty wax into the cappings of the comb honey.

It is common to use queen excluders when producing comb honey, as they keep the queen out of the honey supers and reduce the amount of pollen stored above the brood nest.

Further comb honey supers are usually put on above the others (over-supering), unless the honey flow is likely to continue for some time. If the flow is expected to continue well, you can put a comb honey box on below the other partly full ones (this is called under-supering). But if you do this and the flow cuts out shortly afterwards, a lot more drawn out and partly-filled combs will be left. Be conservative when supering up for comb honey production. The bees need to be crowded in the hive and forced onto the foundation to produce a good product.

Because comb honey hives are crowded they need extra ventilation, so chock the bottom brood box up off the floorboard during hot weather, especially if the floorboard risers are only 10–12 mm deep. The honey supers can be staggered but should not be shifted enough to admit bees, or pollen will be stored in the honey frames. Crowding also means these hives are more liable to swarm, so young queens must be used.

REMOVING COMB HONEY

Comb honey should be taken off the hive as soon as it is fully capped; otherwise bees will track propolis or pollen stains onto the surface of the comb, leaving it 'travel stained'. Bees can also empty out some cells in the middle of the comb, which means this part of the comb is not first grade (though it can still be used if you place this surface face down on the bottom of the pack where it is not seen).

Bee escape boards or fume boards are best for harvesting comb honey from the hive. If you are using escape boards on more than one super on a hive, crack all the supers apart before putting the escape board on. This breaks any brace comb between the boxes, so the bees can clean up the dripping honey before the boxes are taken off.

STORING COMB HONEY

Store comb honey for as short a time as possible before packing it, to stop wax moth infestation. One storage method is to stack the boxes of comb honey on drip trays with a sheet of newspaper between each box. This stops wax moth spreading should it infest one box, and also stops honey dripping from broken brace comb over the face of lower combs. Store the stacks where it is warm (around 20 °C) to retard granulation, and dry to prevent moisture uptake.

A safer method is to store comb honey in a freezer until packed. Comb honey can also be kept frozen after packing to stop the honey from granulating.

PACKAGING

Before any comb honey is packed it must be treated to kill wax moth eggs, larvae or adults. Commercial beekeepers usually freeze stacks of supers in commercial cool stores. For smaller-scale beekeepers, honey can be frozen in a domestic deep-freeze for 24–48 hours. Allow the honey to return to room temperature before packing it, but be extremely careful how you store it to prevent wax moth from re-infesting the honey boxes.

Cut comb honey

Before squares or rounds can be cut from the comb, you first have to remove the wires. To do this, lay the frame on a queen excluder with the comb's best face uppermost and cut each wire at both ends. Using insulated pliers connected to a battery terminal or transformer, grip one end of the wire (as with embedding wax foundation), and connect the other end to the opposing terminal. As soon as the wire is melted free of the foundation, disconnect the current and quickly pull the wire out.

Fig. 17.4 Cutting cut-comb honey out of the frame.

The next step is to cut pieces from the comb that will fit neatly inside plastic comb honey boxes, which can be bought from beekeeping equipment stockists. The easiest way of doing this without building a special steam or electrically heated cutter is to cut measured squares of comb with a cold, sharp kitchen knife. If building a cutter, make it out of copper and have it tinned (Fig. 17.4). Stainless steel does not transmit heat readily enough, even though it is the preferred material for contact with honey.

Fig. 17.5 Draining cut-comb honey.

Lift each square gently off the queen excluder (Fig. 17.5) with a fish slice, and place it in a plastic comb honey box. Wipe off any excess honey with a damp cloth, before putting the lid on and sealing the joint with clear tape. Cut-comb squares intended for sale

Fig. 17.6 Square and round packages of cut comb honey.

should be labelled with the name of the product, 'comb honey'; the name and address of the producer or vendor; a batch code or date; a nutrition panel and a minimum net weight. This is usually expressed in a form such as 'not less than 340 g'.

Some hobbyists pack cut comb in plastic bags, plastic or styrofoam trays, or even pieces of plastic cling film. This is just adequate for home use or gifts, but not for sale. The honey is easily crushed in such flimsy packaging, and doesn't look attractive to potential buyers.

Packing sections

Press the sections out of the frames onto a flat bench top kept clear of any propolis or wax that might damage the comb surface. Scrape the propolis from the wooden section with a knife, or buff the wood with a rotary wire brush mounted on a horizontal spindle (e.g. on a bench grinder).

When the section is cleaned, wrap it in a cellophane wrapper, which you can buy from some bee equipment stockists. Fold the bag neatly with a parcel fold and seal it with an electric iron set on a moderate heat. Zip-lock or sealable plastic bags will also do, but are not as attractive as a cellophane wrap.

Sections to be sold must be labelled in the same way as cut comb honey.

Chunk honey

Chunk honey is rarely seen in New Zealand, but is very popular in the United States and Europe. It consists of thin pieces of comb honey, perhaps offcuts from cut comb, placed in a jar which is then filled with liquid honey.

The problem with chunk honey is that the liquid honey can granulate quite rapidly, leaving an unsightly jar of coarsely granulated honey. The liquid honey used to fill jars of commercial chunk honey is usually light coloured and of a sort that is slow to granulate naturally (such as viper's bugloss or thistle), or other types such as clover that have been processed to retard granulation. Some beekeepers may flash-heat the honey first to slow down the onset of granulation, but even so the product has a limited shelf life of only three to six months.

To produce chunk honey for home use, pack it in jars then store it in a freezer. Granulation is very slow at freezer temperatures, and jars can be returned to room temperature and used as required. Once removed from the freezer, the honey has a short shelf life before it granulates.

UNFINISHED COMBS

Partially drawn-out sections or cut comb frames can be cleaned by bees inside a super over a hive, and can be used again next year provided the wax is not too stained. The offcuts from cut comb processing should be treated in the same way as cappings from extracted honey, which are discussed in the next chapter.

Beeswax <18>

Beeswax is a wonderful hive product. Beekeepers use it themselves as new comb foundation or for waxing plastic frames, and making pure beeswax candles is a pleasurable sideline. Around 250 tonnes of beeswax is collected by beekeepers in New Zealand each year, nearly half of which is exported. In this country beeswax is used mainly by the beekeeping industry itself. Cosmetic and polish manufacturers also use large quantities, and there are literally hundreds of minor uses — waxing fabrics and threads, candle making and in dentistry, for example.

Beeswax is a secretion of worker honey bees, produced by four pairs of glands on the underside of their abdomens. Bees use this wax to make combs for food storage and the rearing of brood.

PHYSICAL PROPERTIES

Beeswax is always white when first secreted, regardless of the food consumed by the bees producing it. Wax usually takes on a yellow colour while in the hive as pigments from pollen are incorporated. Brood-rearing darkens a wax comb even after only one cycle of brood has been raised. This coloration comes from the used pupal cocoons and accumulated larval faeces in the cells, and brood combs darken further with time until they are virtually black.

Beeswax is a very stable substance, and its composition does not alter significantly with storage. The melting point of beeswax is 64 °C, the highest for any known natural wax. Its specific gravity of 0.96 means that it floats on water, which is very important for many processing methods.

SOURCES OF BEESWAX

There are three main sources of beeswax for a beekeeper — cappings, scrapings, and old or damaged combs culled from the hive. Wax from old brood combs should be kept separate from other types of wax, as it is usually very dark. Lighter-coloured wax is worth more than dark wax, and mixing different coloured waxes reduces the overall value of the product.

Fig. 18.1 Cappings are the beekeeper's most important source of beeswax.

Fig. 18.2 Removing burr comb from top bars.

Cappings

The most important source of wax for most bee-keepers is cappings removed from honey combs during processing (Fig. 18.1). Depending on how fully capped the combs are, and how many frames are in each honey super, about 14–18 kg of wax is removed per tonne of honey extracted. Cappings wax is light in colour if removed from 'white' combs (in which no brood has been reared), but darker if it is from combs that have been used for brood rearing.

Scrapings

Beeswax can also be scraped from hive parts. During honey flows, bees deposit considerable quantities of wax (called burr comb) on frames (Fig. 18.2), queen excluders and the insides of hive boxes. If the bee space between frames is not correct, bees also build extra comb (called brace comb).

There are two good reasons for removing this wax. First, if the wax is allowed to accumulate you will find it increasingly difficult to manipulate the frames, and hive management becomes more time consuming. Second, wax should be salvaged from the hive because it is a valuable product.

On average up to half a kilogram of wax can be scraped from a hive during ordinary hive management each year. The small pieces of scrapings wax seem unimportant and are discarded by some beekeepers, but if they are saved a significant amount of wax accumulates. In any case, wax scraped from the hive should not be discarded in the apiary (Fig. 18.3) as it could spread American foulbrood and provoke robbing.

Culled combs

There are several reasons to remove (or cull) combs from a brood nest (Fig. 18.4). Beekeepers used to cull combs that contained a significant portion of drone comb, which is made by bees when they repair broken parts of the frame. But there is no need to cull combs that have a small proportion of drone cells. In fact, with the advent of plastic frames and the varroa mite (which selectively attacks drone larvae), many apiaries may now have too few drones to ensure that queen bees are adequately mated. As a result, beekeepers now cull out frames with drone comb less rigorously than they used to, and more progressive beekeepers even insert drone foundation frames into the brood nest of their hives. On the other hand, some beekeepers encourage drone comb production, and cull these frames when full of drone brood as a means of controlling varroa.

Brood combs clogged with old pollen leave no room for the queen to lay eggs, and should be culled. Old brood combs also act as a reservoir for nosema spores, and regular comb replacement may be an important part of nosema control where this disease is a problem.

Continued brood-rearing in a comb may reduce the size of cells and thus of the bees reared in them, but the significance of this for honey production is not known. Some beekeepers advocate the use of small cells to restrict the development of varroa, but this has not been proven to be effective in New Zealand.

As a rough guide, you could expect to replace around three combs per year in established hives, either by removing the whole frame and melting out the wax, or in the case of plastic frames by scraping the wax off with a hive tool, paint scraper or water blaster.

Recovery of wax from these culled combs depends on the efficiency of the processing system, but yields on average 1.5 kg per 10 full-depth frames. Slightly higher figures can be achieved when processing very heavy combs with thick foundation. The value of wax recovered from old combs exceeds the cost of fitting replacement foundation, so even when processing costs are taken into account, a regular comb replacement programme does not need to be expensive.

Beeswax must not be rendered out from combs infected with American foulbrood. Under the AFB pest management strategy, diseased combs must be burnt and the remains buried.

PROCESSING CAPPINGS AND SCRAPINGS

Draining cappings

Cappings consist of a mixture of wax and honey, and the honey included with the cappings can be more valuable than the wax. The main aim of processing

Fig. 18.3 Discarding wax in the apiary is wasteful and a disease risk.

Fig. 18.4 Combs being culled from the hive.

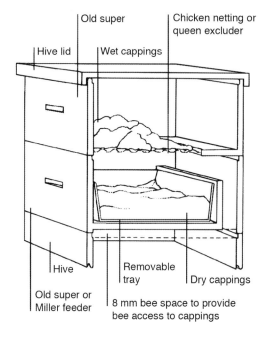

Hive lid

Old super

Wet cappings

Chicken netting or queen excluder

Hive

Removable tray

Dry cappings

Old super or Miller feeder

8 mm bee space to provide bee access to cappings

Fig. 18.5 Equipment for drying out cappings on a small scale.

cappings should be to remove as much of that honey as possible without damaging it by overheating.

Small-scale beekeepers normally do this by draining. A simple method is to uncap into a super with a queen excluder screwed to the bottom, or a draining basket made of gauze or lined with cheesecloth or nylon mesh. Cappings can also be hung up in a draining bag. Put the cappings in a warm, bee-proof place for several days, and stir occasionally to help drainage. The honey that drains off is perfectly suitable for mixing back with the rest of the crop.

Not all the honey will drain out. Some beekeepers wash out the remainder and make mead with the honey/water mixture, but it is more common to feed the honey directly back to the bees. Do this inside the hive, and put the honey out at night. Never feed honey in the open, as this can cause fierce robbing. Feeding extracted honey can also spread AFB.

Cappings honey can be fed back to the bees in several hives close to home — there is no need to spread it over all your hives. A simple method is to put wet cappings into a large dish or pan inside an empty super on a hive. Turn the cappings occasionally to give bees access to the honey.

A more convenient way is to make a simple device for holding the cappings while they are being dried (Fig. 18.5). This can easily be made from two old supers, or wood cut to a similar size. Put wet cappings in the top half. Bees gain access to them through the chicken netting or queen excluder, and as honey is removed the dry cappings fall down into the removable tray below. You don't need to make visits to turn the wet cappings, and dry cappings are simply removed in the tray.

Colonies fed wet cappings can be left with less honey in the brood nest, as their winter stores will be made up by the honey removed from the cappings. This method can stimulate late brood rearing, but is the easiest way of drying small quantities of cappings.

Melting dry cappings and scrapings
Solar wax melter

By far the easiest method of rendering small quantities of cappings and scrapings is to use a solar wax melter (Fig. 18.6). This simple piece of equipment consists of a wooden box containing a metal tray on which the unprocessed wax sits, and a small mould into which the molten wax runs. A glass lid retains the heat, and the box is tilted at an angle to catch the sun. The size of a solar wax melter is not important as long as the basic principles are followed.

A solar wax melter has five parts:

Fig. 18.6 Solar wax melter.

- a wooden body
- a glass or clear plastic lid
- a large metal tray on which the crude wax is placed
- wire mesh to strain the molten wax
- a small pan in which to catch the molten wax.

The wooden body can be made from 20 mm timber or plywood. Glue and screw all the joints to prevent buckling with the intense heat that will be produced. Paint the body black on the outside (to absorb heat) and white on the inside (to radiate heat).

The glass lid can be based on an old window frame or built from scratch. In either case it is best double-glazed, with two sheets of glass separated by a 10–15 mm gap or one sheet of glass and one of clear plastic. If condensation in this gap is a problem, put in a small quantity of silica gel (obtainable from pharmacies).

The metal tray can be made of flat or corrugated galvanised steel, aluminium, tin or even stainless steel. Do not use non-galvanised steel, zinc, brass or copper, as these can discolour the wax. The metal should be formed into a tray with one or two spouts at the lower end. A wire-mesh lining will filter out impurities or debris in the wax. Finally, a mould must be provided for collecting the molten wax. Plastic or any of the metals that are suitable for the tray can be used.

Place the solar melter in a sheltered, sunny position, since wind greatly reduces its efficiency. If the lid does not fit snugly, seal any gaps with strips of foam plastic.

A solar wax melter is cheap and easy to make, and costs nothing to run. It can be used for all types of wax, and a big advantage is that you can melt wax as it comes to hand rather than leaving it to be processed in the off-season. This is useful as wax left lying about, particularly in the form of old combs, is soon destroyed by wax moths. Solar melters also bleach the wax, so improving darker comb wax. In some parts of the country though, solar wax melters cannot be used all year round.

Hot water

Cappings that have been dried out over a hive and are relatively free of honey can be melted in hot water. This must be done carefully because of the fire danger from hot wax.

A large container like a preserving pan is needed, but remember that once used to melt wax, it cannot be used for food again. Fill it about one-third full with hot water (no hotter than 90 °C) and add the cappings slowly so the water does not cool down too much. On no account boil the water as this emulsifies the wax and incorporates impurities, and wax spatterings increase the already considerable fire danger and mess.

When all the cappings have been melted, leave the mixture to stand until a fine film of wax forms on the surface, before pouring it into moulds. This resting period allows dirt to settle and ensures that the wax is not hot enough to melt plastic moulds.

A mould can be made of stainless steel, tin, plastic, aluminium or galvanised steel, and preferably with tapered sides. If the sides do not taper, slightly constrict the top of each container by tying a piece of string tightly around it, after placing a small wooden block on each face. The blocks depress the sides slightly, so that when the wax has set in the mould and you remove the string and blocks, the sides will spring out slightly. This will admit air to release the suction, and allow the wax block to be removed.

Cover moulds with cloths, sacks or polystyrene sheets while the wax is cooling, so that it sets slowly and firmly into solid blocks. If the wax cools rapidly the blocks will crack.

Melting cappings and honey

You can melt cappings before they are drained, but because the temperatures needed to melt wax also damage honey, this approach is less satisfactory than processing cappings that have been drained of honey.

Oven

You can melt undrained cappings in a domestic oven set at its lowest temperature. They should be placed in a mould such as a stainless steel bowl, heated gently and stirred occasionally. When the wax has melted, switch off the oven and leave the wax to cool. The remaining honey at the bottom of the block of wax is usually suitable only for baking.

Water bath

A wax/honey mixture can also be melted in a container heated in a water bath. When the wax has melted, remove the container from the hot water, cover it with a heavy cloth and leave the wax to set on top of the honey/water mixture in the bottom.

Cleaning queen excluders

Queen excluders are difficult to scrape clean without bending the wires, so wax on them is best removed by putting the whole excluder in a solar wax melter for several hours.

PROCESSING OLD COMBS

Dealing with old combs can be a problem for hobbyist beekeepers, as large and expensive equipment is necessary to do the job effectively. Because of this, many domestic beekeepers fail to replace combs as often as they should, or simply throw away those that are culled. There are, though, fairly efficient ways that are suitable for hobbyists to get wax from old combs.

Fig. 18.7 Commercial hot-water melter for rendering beeswax.

Solar wax melter

An efficient solar wax melter will recover up to half of the wax in old combs, provided they are laid out in the melter in a single layer. Any remaining wax is bound up in the 'slumgum', a mass of old pupal cocoons, pollen and general debris from the comb. It is not economic to process small quantities of slumgum from a solar wax melter, and this is best spread on the garden since it makes an excellent mulch and fertiliser. Large amounts of slumgum can be processed by a beekeeper with an efficient wax plant or by a beeswax processing firm, but even this may be only marginally profitable.

Hot-water wax melter

A hot-water wax melter can deal with beeswax from old combs and other sources (Fig. 18.7). A simple version of such a melter (Fig. 18.8) is fairly cheap to make, and can be shared by several beekeepers or a beekeeping club. It has an electric immersion element to heat water, and combs are melted in the hot water. Once it is full, a mesh screen is placed over the wax/water mixture. Tipping a quantity of hot water into the funnel inlet forces the

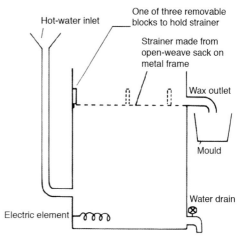

Fig. 18.8 Simple hot-water melter for rendering beeswax.

Fig. 18.9 Blocks of beeswax ready for sale, conversion to comb foundation or waxing plastic foundation.

wax up through the screen (which filters impurities) and out into a mould.

DISPOSAL OF BEESWAX

Light and dark wax should be processed separately, since the price of wax depends on its colour and mixing colour grades will lower overall returns.

Blocks of crude beeswax (Fig. 18.9) can be sold to beeswax merchants. They can also be sent to beeswax processors and swapped for the equivalent weight in comb foundation. A conversion fee and freight will be charged, but this works out cheaper than selling wax and then buying foundation with the money obtained.

COMB FOUNDATION

Foundation can be bought from beekeeping equipment suppliers in lots of 10 sheets or, at a considerable saving, in cartons. A 15 kg carton of medium brood foundation contains about 315 three-quarter-depth sheets or 255 full-depth sheets. Heavier-grade wax sheets may come in 17 or 20 kg cartons.

Store foundation until it is needed in a cool place away from light and all insecticides, especially pest strips and fly spray, which are readily absorbed by the wax and may kill bees when the foundation is used later.

Before embedding foundation in the frame wires, remove it from storage and leave it at room temperature (around 20 °C) for about 24 hours. Cold foundation is brittle and will break if not handled carefully.

Comb foundation that has been stored at low temperatures may develop a surface bloom, which looks like mildew or a fungal growth. This is simply a physical property of the wax and is harmless to bees. Foundation can be stored for years and is not usually attacked by wax moths.

USES FOR BEESWAX

The section on further sources of information at the end of this book lists two titles that deal with making beeswax candles, and there is plenty of information available on the internet.

There are many different recipes for making polishes from beeswax, and we give four here. There is a high fire risk in making wax polishes, so be especially careful not to spill molten wax on any solvents or onto a direct heat source.

FURNITURE POLISH

Ingredients
500 g beeswax
85 g potassium bicarbonate
2.25 litres water
1 litre turpentine

Method
Grate the beeswax and heat with the water and potassium bicarbonate until melted. Pour into a large enamel or stainless steel vessel well away from the heat source, add the turpentine and stir vigorously with an electric beater or paint stirrer until well emulsified. Pour into suitable containers. Colouring such as yellow ochre or shoe polish can be added if desired.

Ingredients
600 ml linseed oil
1.2 litres turpentine
60 g beeswax

Method
Melt the beeswax in warm linseed oil, and then add the turpentine and mix. Shake well before use.

FLOOR POLISH

Ingredients
500 g beeswax
2 litres boiling water
2.5 litres turpentine
100 ml cloudy ammonia

Method
Melt the beeswax slowly, take off the heat and add the boiling water, turpentine and ammonia. Stir together, and stir again occasionally until cold.

Ingredients
250 g beeswax
300 ml turpentine
125 ml alcohol

Method
Melt the wax in a double boiler, and then add the turpentine and alcohol. Stir the mixture until it forms a thick paste and pour into containers.

19 Pollination

Beekeeping plays a very important part in the New Zealand economy. Many of New Zealand's fruit and pastoral seed crops rely on cross-pollination for an economic level of seed set or fruit development. Many crops are insect pollinated, and by far the majority of insect pollination is done by honey bees. The direct value of honey bees to the country as a result of their pollination has been estimated at over 60 times the value of the products and services they otherwise produce.

The advent of varroa in New Zealand has significantly reduced the numbers of feral honey bee colonies, making it even more important for growers to manage pollination in order to grow many crops profitably. Growers who used to rely on feral bees for pollination are now finding they need to pay for pollination services.

There are several reasons for the honey bee's effectiveness as a pollinating agent:
- Honey bees collect and consume pollen as a source of protein, and in doing so cross-pollinate flowers. Their bodies are specially adapted to collect pollen by being covered with branched hairs, and they have pollen-collecting structures on their legs such as the antenna cleaners, pollen rakes and pollen baskets.
- Honey bees live in very populous colonies that are perennial, i.e. continuing from year to year.
- They are unique among insects because they have an efficient communication system to recruit foragers onto particular food sources, and also because they usually specialise either as nectar or pollen gatherers.
- Honey bees show remarkable constancy in collecting nectar or pollen from a single plant species on any trip from the hive. This ensures that pollen from one flower is not wasted by being transferred to flowers of another species.

Advanced management techniques can be used to increase the efficiency of honey bee colonies for crop pollination, and colonies can be built up in strength and moved

Fig. 19.1 Pollen-collecting bee transferring pollen between flowers.

onto a crop in large numbers at the required time. Honey bee colonies can also be removed from target crops when pollination is completed, to prevent them from being poisoned by insecticides, and to stop them from interfering with other orchard or farm activities.

Because honey bee colonies are so populous, they can each perform a vast amount of work. In a pollination hive of 30,000 worker bees, about 12,000 will be foragers. Gathering enough nectar to

Fig. 19.2 Pallets of hives in a kiwifruit orchard.

make 1 kg of honey will involve the bees in a total of about 60,000 flights from the hive. When pollinating kiwifruit, for example, an individual bee may visit over 100 flowers in an hour. A honey bee colony of average size will consume up to 40 kg of pollen per year, all of which they collect from their environment.

USING HIVES FOR POLLINATION

Many people become interested in beekeeping through their involvement with horticulture and the need to pollinate crops. Remember though, there is no such thing as free pollination. Running your own hives simply to save the cost of renting hives at blossom time can be an expensive way of pollinating your crop.

If pollination is your sole aim, think carefully about the following points before buying beehives, as it may be cheaper (and a lot easier) to use the services of a commercial beekeeper:

- The hives must be located off the property all year round except when they are brought in at blossom time. This is to keep them safe from possible insecticide poisoning, and because bees in hives brought in do more pollinating than bees in hives permanently sited there. When a hive is moved to a new site, the bees go out and search actively for food sources. Known as 'naive foragers', these bees do a significant amount of pollination. Permanent hives establish flight paths to other food sources, and may be slow to move onto the target crop when it begins to flower.
- Beekeeping management usually conflicts with a grower's work schedule. Honey bee colonies must be managed intensively, especially during spring, to ensure that they are strong units at flowering time. This is particularly true given the advent of varroa in New Zealand. Growers and farmers are often busy in spring, and so hive management may be neglected at a critical time.
- Many people who own hives mainly for 'free' pollination lose interest in managing the bees, especially after a few episodes of bad stinging. This can begin a vicious cycle of poor management leading to swarming and a more aggressive strain of bee, resulting in even less interest in the hives, and so on. It often culminates in the hives being destroyed because they have become diseased or simply neglected, but in the meantime the grower's own crop has not been pollinated adequately. Diseased hives put other hives brought in for pollination at risk. If those hives become diseased, pollination beekeepers are unlikely to bring their hives back to that location the next year, and every orchardist or seed farmer suffers.

If you wish to use your bees for pollinating crops, recognise that this requires specialised management, hard work and possibly a reduced honey crop — and is in no way free.

A good pollinating colony for most crops consists of two brood boxes, containing about a box and a half of bees. Single-box hives and nucleus hives are generally not adequate, and are not worth moving into orchards for pollination.

The most effective pollinators in the hive are usually pollen-collecting bees, and the stimulation to collect pollen comes mainly from the presence of unsealed brood.

A pollination hive should have at least the equivalent of six or seven frames each containing a good slab of brood, mainly in the bottom box, and not more than one or two frames of pollen. At least the equivalent of three full-depth frames of honey should be in the hive for stores, to ensure it does not starve during the pollination period.

MOVING BEES TO CROPS

Fig. 19.3 Using a grid to estimate brood area during a kiwifruit pollination hive audit.

The pollination requirements of a crop are often specific to that particular plant. This chapter discusses pollination in general, but professional advice is needed for working out how to optimise pollination in any given situation and with any given crop.

Hives should not be moved into a crop until after pre-blossom insecticides have been applied and the waiting period stipulated on the insecticide label is over. For most crops it is best to move them in at about the 10 per cent blossom stage. If they are shifted in earlier, many bees will become orientated to other food sources and may be slow to work the target crop when it comes into flower. If the hives are shifted in much later, some flowers will have finished their development without being pollinated.

Correct positioning of the hives in the orchard is important, since this can increase the amount of time the bees spend flying. Most foraging takes place when the air temperature is higher than 12–14 °C and when the wind is less than about 25 km/h, so hives should be placed where they are sheltered from prevailing winds and receive maximum sunlight, especially in the morning. Placing hives up off the ground on upturned fruit bins or stands also seems to help get bees flying earlier in the day.

It is much better to choose such sunny, sheltered sites in an orchard, than to spread the hives out to every corner of the property. Orchardists sometimes make this mistake, failing to realise that bees are very good at flying around the orchard and finding flowers with pollen.

Hives should be shifted out from the crop at about the 90–95% flowering stage, and definitely before post-blossom insecticides are applied.

HOW TO MINIMISE BEE POISONING

Some bee deaths are an inevitable consequence of using hives to pollinate fruit crops. However, it is possible to avoid serious losses that will reduce or completely wipe out a colony's production. Beekeepers should move their hives in only after the waiting period is over for pre-blossom insecticides, and remove the hives quickly when pollination is finished.

Growers must also play their part by observing the label requirements on chemicals.

Fig. 19.4 Bees killed by indiscriminate pesticide use.

The law requires that chemicals toxic to bees are not applied to, or allowed to drift onto, any plant that is attractive to bees, either when the plant is flowering or if it will come into flower while the chemical is still toxic to bees.

Insecticides drifting down onto flowers in the orchard ground cover, or onto hedges and shelter belts, are a main cause of bee deaths. It is vital that growers mow orchard grass before using insecticides if flowers are present, and that care is taken not to allow drift onto adjacent plants such as hedges and shelter belts if they are in flower or about to flower.

HIVE REQUIREMENTS FOR PARTICULAR CROPS

The number of honey bee colonies needed to pollinate an area of crop depends mainly on the nature of that crop, and its attractiveness compared with other food sources within flying range (3–5 km).

Each crop has its own characteristics that may make pollination easy or difficult. For instance, kiwifruit produces pollen that bees collect freely, but no nectar. The sugar concentration of pear nectar is often below the threshold of attractiveness for bees, and plum nectar is attractive but the weather is often bad when plum trees flower. Many blueberry crops are grown under enclosures, which restrict bee flight.

The recommendations given in Table 19.1 cover many common crops grown in New Zealand, and are general recommendations for average conditions. For further information contact an apicultural or horticultural specialist, or consult one of the reference titles listed at the end of this book.

Table 19.1   Recommended hive stocking rates
for pollination of fruit and seed crops

Crop	Hives per hectare for mature crops
Fruit crops: berryfruit	
blackberry, boysenberry, raspberry	6
blackcurrant	5
blueberry	1
Fruit crops: pipfruit	
apple	4
pear	6
Fruit crops: stonefruit	
apricot	8
cherry	10
peach and nectarine	5
plum	8
Fruit crops: other	
avocado	at least 4, up to 10 recommended for larger trees
citrus	usually no hives introduced to orchards
kiwifruit, gold	6
kiwifruit, green	8–10
Seed and vegetable crops	
brassicas (e.g. canola)	3
clover	3
cucurbits	2–4
lucerne	3–5
squash	2
sunflower	1–2

20

Laws affecting beekeeping

Whether you are a hobbyist or commercial beekeeper, you do need to be aware of legislation that relates to beekeeping. What you do as a beekeeper affects others — bee diseases will spread to other beekeepers' hives if not managed properly, and honey and other products must be safe for human consumption. Your bees can also cause harm or nuisance to others.

Laws relating to beekeeping aim to protect bee health, ensure the safety of hive products, prevent bees being a nuisance, and foster the industry. Beekeepers also need to know about other laws covering areas such as occupational safety and health, and consumer protection.

The most important pieces of legislation for beekeepers are:
- Biosecurity Act 1993: covers control and prevention of pests and diseases, both endemic and exotic.
- Biosecurity (National American Foulbrood Pest Management Strategy) Order 1998: determines how beekeepers manage American foulbrood disease and requires all apiaries to be registered.
- Biosecurity (American Foulbrood — Apiary and Beekeeper Levy) Order 2003: imposes a levy on all beekeepers and a levy on all apiaries in New Zealand, to fund the pest management strategy.
- Animal Products Act 1999: prescribes the processing, handling and storage of bee products that are to be exported with an official assurance (export certificate).
- Food Act 1981: controls processing, handling and storage of food that is for domestic consumption only, or exported to countries that do not require an official assurance. Note: This Act is currently under review.
- Food Hygiene Regulations 1974: control processing and handling of food for domestic sale only.
- Food (Tutin in Honey) Standard 2010 and Animal Products (Harvest Statement and Tutin Requirements for Export Bee Products) Notice 2010: manage the risk

of the toxin tutin occurring in honey for sale or export.
- Local Government Act 2002: allows the formulation of bylaws covering beekeeping.

The full text of legislation referred to in this chapter, and further information, can be found on-line:
- all New Zealand Acts, bills and regulations at www.legislation.govt.nz
- American foulbrood legislation also at www.afb.org.nz
- explanations of food safety legislation at www.foodsafety.govt.nz
- documents from the industry/government consultative forum Bee Products Standards Council at bpsc.org.nz.

This chapter summarises the main provisions of legislation related to beekeeping, but if you need an authoritative opinion you should consult a lawyer.

KEEPING BEES

Before you locate beehives anywhere, you need to know that:
- you will require the permission of the landowner
- the right to put hives on some land owned by the government or a forestry company may be subject to ballot or tender
- many city and district councils have bylaws controlling beekeeping activities, usually based on a model bylaw but with variations that may include a fee to register your hives and/or a requirement for you to get permission from all your neighbours
- some bylaws do not require you to get permission, provided you follow industry good practice and do not cause a nuisance
- even if a council does not have bylaws that deal specifically with beekeeping, if someone complains about your beehives council staff may investigate and even require you to remove some or all of them, using powers under the Local Government Act 2002 or general nuisance provisions under the Health Act 1956.

You must register any place where you keep bees (an apiary) under the Biosecurity (National American Foulbrood Pest Management Strategy) Order 1998. Under that regulation:
- an apiary is a place where one or more beehives are kept
- a beehive is defined as a thing constructed for the keeping of honey bees, and that is being used or has been used for that purpose
- an apiary must be registered if a beehive has been sited there for more than 30 consecutive days

- if groups of beehives are located on the same property and arc more than 200 metres from each other, then each group must be registered as a separate apiary
- if more than one beekeeper has hives in a place, each beekeeper must register that place as an apiary
- beekeepers are allocated a code that must be displayed on the outside of at least one hive or on a sign in each apiary.

Fig. 20.1 Apiary identified with the beekeeper's code number.

CONTROLLING PESTS AND DISEASES
American foulbrood (AFB)

This was the first bee disease to be regulated in New Zealand, when the Apiaries Act 1907 came into force, and was regulated by the government under subsequent Apiaries Acts (the last dating from 1969) for over 90 years. In October 1998 things changed radically, and AFB control is now carried out under a pest management strategy (PMS) under the Biosecurity Act. The strategy is the responsibility of a management agency, which is the National Beekeepers' Association (NBA).

This change was made by repealing relevant portions of the Apiaries Act 1969, and introducing the Biosecurity (National American Foulbrood Pest Management Strategy) Order 1998. The American foulbrood PMS is funded by a compulsory levy on all beekeepers and apiaries under the Biosecurity (American Foulbrood — Apiary and Beekeeper Levy) Order 2003.

The AFB PMS retained many of the provisions from the previous Apiaries Act 1969, and added some new ones. The NBA contracts individuals and companies to implement various parts of the strategy. More information can be found at www.afb.org.nz and at www.nba.org.nz.

Key features of the AFB PMS are:
- all apiaries must be registered if bees or beekeeping equipment are kept there for more than 30 consecutive days
- all hives must be inspected annually by an 'approved beekeeper', who must also report on the disease status of the hives
- reporting is done in an 'annual disease return' (ADR), which is sent to all registered beekeepers in April each year and must be returned to the management agency on or before 1 June in the same year (in future this will be done on-line)

- to become 'approved', beekeepers must first pass a competency test on AFB recognition and control, and then submit a hive and AFB management plan (called a 'disease elimination conformity agreement' or DECA) to the management agency

- DECAs are reviewed frequently but reissued only if major revisions are needed

Fig. 20.2 Destruction of hives infected with American foulbrood.

- DECAs may be revoked if beekeepers are in serious default of their management plan or other aspects of the strategy
- beekeepers who are not 'approved beekeepers' must engage an 'approved beekeeper' or 'authorised person' to inspect their hives for AFB between 1 August and 30 November each year, and report within 14 days of making the inspection or by 15 December at the latest
- an 'authorised person' is an appropriately qualified person appointed by the government, who is given powers, duties and functions in connection with the pest management strategy
- any case of AFB must be reported to the management agency within seven days of its discovery
- any case of AFB must be destroyed within seven days of its discovery unless other options are allowed for in the beekeeper's DECA, such as storing the affected hives in a bee-proof environment until fire bans are lifted, or if an 'authorised person' directs otherwise
- beekeepers must submit samples of bees (hobby beekeepers) and/or honey (commercial beekeepers) to a named laboratory for AFB testing if so requested
- all hives with AFB symptoms must be destroyed, although some equipment can be sterilised by heating in paraffin wax at 160 °C for at least 10 minutes
- antibiotics cannot be used to control AFB in New Zealand.

The AFB pest management strategy is funded by an annual levy on beekeepers and apiaries under the Biosecurity (American Foulbrood — Apiary and Beekeeper Levy) Order 2003. Under this order:

- all beekeepers are required to pay a base fee of NZ$20 per beekeeper plus a fee per apiary
- the apiary portion of the levy is currently (2011) set at $12 plus GST but is reviewed annually by the management agency, and the maximum under the law is $15.17 plus GST per apiary

- beekeepers with fewer than four apiaries *and* fewer than 11 hives pay the base fee plus one apiary fee
- beekeepers above the thresholds are levied the base fee plus the set fee for each apiary
- the levy is payable on all apiaries registered at 31 March, whether or not there are live hives in the apiary.

Varroa

Varroa is now deregulated, as all regulations and *Gazette* notices controlling this pest were removed in September 2008. However, as some islands such as Stewart Island and the Chatham Islands have not reported varroa, beekeepers should be cautious about taking bees or used beekeeping equipment there.

Importation of bees, beekeeping products and beekeeping equipment

The importation of live bees and semen, bee products and beekeeping equipment is very strictly controlled under the Biosecurity Act. Permits with restrictive conditions are required for the importation of bee semen, while used beekeeping equipment and protective clothing usually may not be imported.

If you enter the country with any bee product, something containing a bee product, or anything that has been used in connection with beekeeping (including clothing), you must declare it in your passenger arrival form.

Conditions for the importation of specified processed bee products are set out in an import health standard that is available at www.biosecurity.govt.nz

This standard allows imports of bee products from a few selected Pacific Island countries that have been assessed as free of European foulbrood disease, as well as other products where the amount of honey is limited or the product has been heat-sterilised. Some products require import permits in advance from MAF Biosecurity New Zealand, while others are given a biosecurity clearance or direction at the border. Further work is being done on a risk assessment and import health standard for imported bee products. Contact MAF Biosecurity New Zealand for current information.

Use of drugs

The Agricultural Compounds and Veterinary Medicines Act 1997 is administered by the New Zealand Food Safety Authority (NZFSA), now part of the Ministry of Agriculture and Forestry (MAF). This act addresses risks related to trade, animal welfare, biosecurity and residues in food from chemicals used as agricultural compounds or veterinary medicines.

Compounds have been registered for the control of the mite pest varroa, but no drugs or compounds are registered in New Zealand to control diseases in honey bees except for the use of fumagillin (Fumidil B®) to treat nosema disease in queen-rearing colonies.

PRODUCING, PROCESSING AND SELLING FOOD

It is important that honey and bee products are safe for people to eat, and free of residues that might damage the reputation of hive products. Most of the following regulations apply only if you sell honey, but note that exchanging products for other goods or services (bartering) counts as sale. Even if you produce honey only for your own consumption or as gifts, you should follow good practice to ensure the product is safe and healthy.

Bee products not needing an export assurance

Bee products being produced and processed for sale in New Zealand only, or to be exported to countries that do not require an export assurance (that means countries for which there are no specific import requirements notified under the Animal Products Act 1999), can be handled in one of three ways:

- Processed in premises that are registered with, and inspected by, local authority Environmental Health Officers for compliance with the Food Hygiene Regulations 1974; or

- Produced and processed by someone operating an approved 'food safety programme' under the Food Act 1981. Beekeepers or processors operating a food safety programme are not required to register with, or be inspected by, the local council. Instead they register with the New Zealand Food Safety Authority and are audited at least once a year by an independent auditor approved by the authority; or

- Managed under a new programme being developed called a 'food control plan' (FCP), that will come under the new Food Act currently being drafted. Beekeepers choosing to operate under an FCP will develop a set of procedures setting out how they actively manage food safety. These can be downloaded as 'off the peg' plans from the New Zealand Food Safety Authority website (www.foodsafety.govt.nz), or operators can prepare their own. Operators working under an FCP will be exempt from the Food Hygiene Regulations. The plans will be registered with the New Zealand Food Safety Authority, and an Environmental Health Officer (or representative) or other accredited verifier will check that the business is following an appropriate FCP by reviewing records, talking to management and staff, and assessing business activities.

Bee products needing an export assurance

Bee products that are produced and processed for export, and for which official assurances (export certificates) are required or requested, must be processed under the requirements of the Animal Products Act 1999.

This act requires secondary processors (those who extract, process, pack or store bee products) to have a risk management programme (RMP) registered with the New Zealand Food Safety Authority. Secondary processing commences once the raw material (e.g.

Fig. 20.3 Honey house compliant with food processing legislation.

pollen, propolis, honeycomb or cells containing honey bee larvae or royal jelly) arrives at the facility where it will be extracted, dried, or otherwise processed, packed or stored. Premises are audited at least once a year by approved verifiers or auditors.

RMPs are designed to identify, control, manage and eliminate or minimise hazards and other risk factors, so that the food is fit for its purpose. In most cases this means fit for human consumption, but the product could also be for animal consumption. Consumption can also mean a bee product you apply to your skin. A code of practice has been developed for the beekeeping industry that covers most processing and storage operations. Operators who adopt this code do not have to have all the procedures evaluated, as these are already approved by the New Zealand Food Safety Authority.

A regulated control scheme to sample and test for residues in honey came into force in April 2009, replacing a voluntary scheme managed by the industry to meet EU requirements. The Animal Products (Regulated Control Scheme — Verification of Contaminants in Bee Products for Export) Notice 2009 requires honey samples to be taken from a selection of producers who operate honey factories under a risk management programme. Operators of RMPs are levied by the New Zealand Food Safety Authority to fund sampling and testing.

Other requirements for bee products sold

If you sell honey you will be subject to the food labelling and composition standards of the Australia New Zealand Food Standards Code 2002. Among other things the food standards code details labelling requirements for retail and bulk packs. Minimum requirements for retail packs of honey are: the name of the product (e.g. 'honey'), a lot or batch code, the packer's name and physical address, any mandatory warning labels (e.g. if the product contains pollen or royal jelly) and a nutrition panel. More information is available at www.foodsafety.govt.nz and www.foodstandards.govt.nz.

Beekeepers who sell bee products to an exporter, or sell on the local market, should also be aware of the Animal Products (Specifications for Products Intended for Human Consumption) Amendment Notice 2009. This notice requires producers to ensure that beehives are made of, and maintained with, materials that do not introduce a hazard into the honey or other bee products harvested from the hive. Beekeepers must also store and transport honey supers, both before and after harvest, in a way that does not introduce contaminants.

Beekeepers carrying out a business activity or offering a service also need to comply with consumer protection legislation such as the Fair Trading Act 1986, the Consumer Guarantees Act 1993 and the Weights and Measures Act 1987 (for further information see www.consumeraffairs.govt.nz).

Toxic honey management

Tutin is a very toxic compound found in the tutu plant (*Coriaria arborea*) (Fig. 20.4). Toxic honeydew (insect excretion) is produced by the passion vine hopper (*Scolypopa australis*) (Figs 20.5 and 20.6), an insect that lives on tutu bushes as well as many other woody plants. Tutu is widespread throughout New Zealand, but this insect lives only in many areas of the North Island and the top part of the South Island.

Fig. 20.4 Tutu, the source of toxic honeydew.

Under certain conditions, bees will collect toxic honeydew from the stems and leaves of the tutu bush and incorporate it into honey — this can occur when there is limited or no nectar, and long dry periods allow the honeydew to accumulate on the plant surfaces. Contaminated honey has no distinctive colour or taste, and tutin is not toxic to bees. Over the years a number of people have become ill after eating honey containing tutin, with a recent occurrence (2008) affecting more than 20 people.

Fig. 20.5 Adult passion vine hopper.

The areas historically regarded as high risk for the production of toxic honey are the Coromandel Peninsula, eastern Bay of Plenty and Marlborough Sounds. Further testing after the 2008 poisoning has identified other areas at risk, including Gisborne, Wairoa and some parts of the Nelson district. Check www.foodsafety.govt.nz for current information.

The Food (Tutin in Honey) Standard

Fig. 20.6 Immature passion vine hopper.

2010 applies to all beekeepers who sell or barter honey, people and businesses selling honey, the last person to pack honey for retail trade and any person who exports honey. The standard applies to honey harvested between 1 January and 30 June (inclusive) in any year. A compliance guide to the Tutin Standard 2010 is available on www.foodsafety. govt.nz.

The standard does not apply to hobby beekeepers who produce honey for their own use or to give away, but does apply if the honey is bartered or exchanged for goods or services. If you are a hobby beekeeper you are encouraged to comply with the standard for your own protection and that of your family and friends.

Under the standard, honey harvested from the beginning of January to the end of June must either be tested for tutin, or the supplier has to confirm that he or she has:

- harvested honey during a low-risk period, which is deemed to be after 1 July and before 31 December; or
- located their beehives in a geographical location that has no significant tutu within the likely foraging range of the bees from those hives; or
- no beehives located in risk areas, which are currently the whole North Island and areas above 42 degrees south (south of Westport on the West Coast and south of Cape Campbell in the South Island); or
- located their hives in risk areas but where the risk of producing tutin honey is low (honey from individual apiaries, or apiaries in close proximity to each other, must be tested over a three-year period before this option can be used).

Section comb honey cannot be tested for tutin before being offered for sale. Cut comb honey can be tested, but if you produce cut comb honey in a risk area (the whole North Island and above 42 degrees south) it is still best practice to harvest it before 1 January each year. Otherwise, if you harvest cut comb honey after 1 January, then you must be able to meet one of the other conditions: no significant tutu bushes within foraging range of the hives; or have apiary and tutu location records plus at least three years of test results from individual apiaries showing either that no tutin was detected in the honey or that tutin levels were under the limits.

Beekeepers who supply honey for export for which an export assurance is required must complete a 'harvest statement' under the Animal Products (Harvest Statement and Tutin Requirements for Export Bee Products) Notice 2009. If you sell honey to a packer, be sure to find out first if they require such a document. The harvest statement requires beekeepers to state if their honey needs to be tested for tutin. If the honey does not need to be tested, the beekeeper must indicate which option from the standard listed above applies to that batch.

You should never eat honey from feral ('wild') bee colonies in the risk areas.

Medicinal claims and nutraceuticals

In New Zealand, foods, medicines and dietary supplements have different regulatory requirements:

- The labelling requirements for food have been outlined above.
- Products claiming to have therapeutic properties are managed under the Medicines Act 1981, administered by Medsafe (part of the Ministry of Health; see www.medsafe.govt.nz). Beekeepers need to be very careful about making any therapeutic claim for their product.
- Dietary supplements are regulated under the Dietary Supplement Regulations 1985, under the provisions of the Food Act 1981. A proposal is being developed to replace this category with 'supplemented foods', to be regulated by a new Supplemented Food Standard. Check www.foodsafety. govt.nz for updated information.

Note also that anyone selling pollen, royal jelly and bee venom must include a health warning advising that the product may cause a severe reaction (see www. foodsafety.govt.nz for details).

OTHER

Pesticides

Pesticides and other hazardous substances are controlled under the Hazardous Substances and New Organisms Act 1996, which is administered by the Environmental Risk Management Authority (www.ermanz.govt.nz).

Pesticides must be registered, and before they are registered they must be thoroughly tested in New Zealand or in recognised facilities overseas. All the proprietary miticides currently allowed for the control of varroa in New Zealand went through this process.

Each year hundreds, if not thousands, of bee colonies are killed or depopulated to some degree by the use of pesticides. The worst damage is experienced on orchards and around seed crops, often not from bees foraging on the target crop but from spray drifting onto flowering plants in the grass cover in the orchards or vineyards. Another cause of bee death is when growers or contractors spray hedgerows or shelter belts that are in flower. Bee losses are also recorded in urban areas where home gardeners spray pesticides onto flowering plants.

Fig. 20.7 Hive floorboard covered with poisoned bees.

The Hazardous Substances and New Organisms Act 1996 and its regulations contain several provisions aimed at protecting bees from incorrect use of pesticides:

- anyone using a pesticide should always read and understand warnings on labels, and abide by the label directions for use
- some pesticides or agrichemicals may require an applicator to be an approved handler and hold relevant qualifications (see www.growsafe.co.nz)
- chemicals toxic to bees must not be applied to, or allowed to drift onto, any plant that is attractive to bees and either is flowering, or will come into flower, while the chemical is still toxic to bees.

Industry development

The two industry bodies, the National Beekeepers' Association (NBA) and the Beekeeping Industry Group (BIG), are funded by voluntary subscription rather than by a commodity levy as in the past. The industry also has an investment fund called the Honey Industry Trust, which is used to promote industry development and support research.

General business legislation

Beekeepers operating a business will have to comply with a range of relevant legislation, covering issues ranging from tax, occupational health and safety, and vehicle usage.

The list of legislation that affects beekeeping is long, and the provisions can seem rather intimidating. But the laws are there for a reason: to protect humans, bees, bee products and the reputation of the beekeeping industry. Use this chapter as a starting point for learning about your responsibilities as a beekeeper.

Beekeeping in New Zealand

21

New Zealand's beekeeping heritage goes back more than 170 years, and today beekeeping is one of the country's important primary industries. Honey, beeswax, bees, propolis, pollen and other minor bee products are exported to dozens of countries around the world. Within New Zealand, honey is enjoyed as a valuable food in its own right, as well as being a useful substitute for imported sugar and sugar-based products.

Many people outside the beekeeping industry associate bees first and foremost with honey production. But honey bees are most important for their role in pollinating New Zealand's horticultural and agricultural crops and pasture legumes, which are the basis for industries worth billions of dollars annually to the New Zealand economy.

SOME HISTORY
Early beginnings

Beekeeping was unknown in New Zealand before European settlement. The first recorded importation of honey bees (*Apis mellifera*) was in 1839. On 13 March that year a mission-ary's sister, Mary Bumby, arrived at the Hokianga in Northland bringing with her from England two bee colonies in straw skeps. The first recorded shipment of bees to reach the South Island arrived in Nelson from Australia on 18 April 1842, consigned to Captain Arthur Wakefield of the new Nelson settlement.

Many other importations followed and beekeeping soon became a popular pastime. In 1848 William Cotton wrote the country's first beekeeping book, *A manual for New Zealand beekeepers*. By the 1860s feral (wild) colonies were common in the bush (native forest), taking advantage of the rich floral sources that were underutilised by other insects and birds.

The first imported bees were of the European or black race. Many other types followed, including what were described as Syrian, Carniolan, Cyprian, Holy Land and Swiss Alpine. Italian queens were first imported in 1880, and the increasing use of this race has led to the present situation where commercial stock in New Zealand is predomin-antly Italian.

No importation of bees has been permitted since the 1950s, except for several quarantined introductions of honey bee semen. Italian honey bee semen was imported in the early 1990s to improve the characteristics of New Zealand's bee stock. Semen from Carniolan stock was also imported in 2004 from Austria and Germany, and used to inseminate selected Italian queens. Naturally mated hybrids from these crosses were released to beekeepers a few months later.

Further importations of semen, coupled with back-crossing, has created an almost pure strain of Carniolan bee in New Zealand that is available to the beekeeping industry. The Carniolan strain of bee is dark in colour with a reputation for being hardy, frugal and more tolerant of varroa mites. However, some crosses can be aggressive and this strain probably swarms more readily than Italian stock. Carniolan bees are becoming more common in the hives of South Island beekeepers, after the removal of all movement controls in 2008 following the spread of varroa outside the controlled areas.

The basis for an industry

All early beekeeping was carried out in fixed-comb hives, either woven skeps (the traditional 'beehive' shape) or box hives. In 1876 the first movable-frame hives were imported and in 1879 Isaac Hopkins, who pioneered many advances in beekeeping, built the first Langstroth hives in New Zealand. This type of hive is used throughout New Zealand today.

Hopkins started keeping bees in 1874, and he was responsible for establishing apiculture in New Zealand on a commercial basis — with Langstroth hives, Italian bees, comb foundation and large, reversible honey extractors. He published a book entitled *The illustrated Australasian bee manual* in 1881, and two years later founded the first beekeeping journal in the southern hemisphere — *The New Zealand and Australian Bee Journal*. In 1884 Hopkins was active in setting up the New Zealand Beekeepers' Association.

During the 1880s the bee disease American foulbrood (AFB) became rampant in the many box hives in New Zealand. For the rest of the 19th century and the early years of the 20th century beekeeping declined, as AFB increased because of the lack of movable-frame hives and beekeepers' ignorance of disease control methods.

As early as 1888 Isaac Hopkins campaigned for bee disease control legislation, but it was not until a year after his appointment as the first Government Apiarist in 1905 that the Apiaries Act, which made box hives illegal, was passed into law. At the beginning of 1908 two government apiary inspectors were appointed: W. B. Bray for the South Island and R. Gibb for the North Island. They travelled by train between districts, taking bicycles with them. Their main task was hive inspection, but they also addressed beekeepers' meetings.

When these first inspectors left to take up commercial beekeeping, four more were appointed, two in each island. Travelling by motorcycle, they inspected hives for American foulbrood and ordered the destruction of box hives.

Another of Hopkins' achievements was to establish an apiary and model bee farm at the Ruakura Government Farm near Hamilton in 1905. This unit demonstrated modern methods of beekeeping to thousands and also trained beekeeping cadets. Many returned servicemen learnt beekeeping there after the First World War. An outstation for queen raising existed at Waerenga near Te Kauwhata from 1908–13.

The first New Zealand beekeeping statistics, published in the national census of 1906, revealed a total of 15,396 beekeepers with 74,341 hives producing 456 tonnes of honey in the course of the year.

Fig. 21.1 Demonstrating beekeeping at the national exhibition in 1907.

The Apiaries Amendment Act in 1913 introduced compulsory registration of hives. By the end of the decade, over 50,000 hives were registered in the names of nearly 5000 beekeepers. Calls by the beekeeping industry during the First World War to increase the number of apiary inspectors were answered by the appointment of some beekeepers as local part-time inspectors. 'Whilst under the present unsettled conditions the Department is not able to offer any salary, they do pay out-of-pocket expenses,' one advertisement read.

Compulsory export honey grading regulations came into force in 1915 to prevent the export of inferior quality honey. Just over 140 tonnes of honey were exported in 1918.

Expansion and development

After the First World War the industry expanded rapidly, as land development increased and more returned soldiers trained as beekeepers. By the end of the 1920s the number of hives had doubled to nearly 100,000, and honey exports had risen to over 1000 tonnes per year.

The 1920s were eventful years for the development of industry organisations. In 1924, for example, the Honey Export Control Act was passed and a Honey Control Board elected. Almost all export

Fig. 21.2 Demonstration apiary at government agricultural research station, Ruakura, 1923.

honey was handled by the Honey Producers' Association, and controlled marketing of New Zealand's honey overseas was a feature of this period.

Fig. 21.3 Commercial apiary in the 1940s.

The depression of the 1930s affected the beekeeping industry along with the rest of the economy. Honey prices dropped to about four pence per pound (8c/kg), and the Honey Producers' Association went bankrupt. In 1934 several honey producers formed New Zealand Honey Ltd to stabilise prices over good and bad seasons, but it failed to achieve its aims because its funds were insufficient to buy all the honey offered in good seasons and it was unable to compete with high prices offered in bad years. In 1938 the assets of New Zealand Honey Ltd were bought by the Internal Marketing Division (IMD) of the government's Marketing Department. The IMD took control of honey marketing, introducing a system of pooling returns from local and overseas sales and paying honey suppliers according to quality. There was no obligation for beekeepers to supply honey to the IMD.

During the depression, retrenchments affected the Department of Agriculture's inspection programmes, and beekeepers were again given authority to act as part-time inspectors. Circumstances later improved and by 1940 the number of hives in New Zealand had risen to more than 122,000, with an estimated annual honey production of over 3100 tonnes. At this time the department employed seven permanent apiary instructors and one honey grader.

Beekeeping increased again after the Second World War, and by 1950 around 7000 beekeepers kept over 150,000 hives. Average annual honey production was calculated at 5000 tonnes. The Department of Agriculture employed 10 apiary instructors.

In the 1950s government policy encouraged the handing over of marketing responsibilities to producer boards, which were set up under the Primary Products Marketing Act 1953. The Honey Marketing Authority (HMA) was set up under this Act in 1953 as a statutory producer board with marketing powers. The IMD was dissolved and the HMA took over the IMD Honey Section's assets, operations and marketing policy.

The HMA remained as virtually the sole exporter of bulk extracted floral honey for the next 25 years, though producers could export their own comb honey, honeydew honey and packed floral honey if they so wished. The HMA bought large amounts of bulk-extracted honey for the domestic and export markets. It was not legally bound to accept all honey sent to it, but did so as a matter of policy in order to stabilise the internal honey market. Honey was graded by government honey graders and payment was based on colour, flavour and quality. The HMA exported most of its honey to a single agent in the United Kingdom.

In the 1970s, following pressure from a number of commercial beekeepers, the HMA began to allow private exports of bulk floral honey. The HMA ceased operations in 1982 and sold most of its assets to a beekeepers' co-operative based in Timaru, which still operates. The ending of controls on private exports of bulk floral honey was an early part of a wider programme of deregulation of agricultural industries in New Zealand.

The Ministry of Agriculture and Fisheries began appointing university graduates as district apicultural advisory officers from 1977, to put beekeeping on the same footing as other primary industries. This change led to the phasing out of the apiary instructor group.

Production and growth

During the 1970s and 1980s, beekeeping enjoyed a growth rate matched by few other primary industries, with a 60 per cent increase in hive numbers over that period. Much of this was as a result of the rapid development of the kiwifruit industry, as well as increased prices for honey. Many people entered the industry by starting new commercial beekeeping businesses, often to provide pollination services to orchardists. The increased demand for such services changed the industry more than anything else up to that time, with most commercial beekeepers involved in pollination to some degree and some even producing little or no honey.

The average size of businesses increased over this time, too, with single-operator enterprises (perhaps with one seasonal worker) increasing from about 600 to at least 1000 hives. There was more diversification in beekeeping, away from the production of bulk extracted honey towards niche marketing of unifloral honeys, more exporting of retail packs, development of fruit and honey spreads, increased volumes of comb honey for export, honey powder used in the food industry, organic honey, dried honey, and the collection and processing of pollen and propolis. Exports of queen bees and package bees grew because of overseas demand and New Zealand's good bee health record.

In recent years the antioxidant, anti-inflammatory and antibacterial properties of honey have been recognised, especially for manuka honey. This has helped turn manuka honey from a product with limited appeal to many beekeepers' highest-value crop.

Difficult times

The 1990s saw a consolidation in the commercial sector, but the change in government policy to user pays affected the industry in a very radical way. Initially the beekeeping industry had to pay for the apiary registration database and some extension and research services. The industry was represented by the National Beekeepers' Association (NBA), which raised funds through a compulsory apiary levy under the Commodity Levies Act 1990.

In 1993 Parliament passed the Biosecurity Act, which consolidated many powers held under a large number of other Acts and regulations including the Apiaries Act 1969. It also removed the responsibility of government for controlling pests and diseases it did not regard as being of national significance, and American foulbrood was treated in this way. The Biosecurity Act allowed government agencies and regional councils to survey for and respond to named pests and diseases, many of which were exotic. The Act also allowed organisations that wished to control endemic diseases to do so under a pest management strategy to be approved by the Minister of Agriculture. The beekeeping industry was one of only two industries to develop a national pest management strategy.

The NBA prepared an AFB pest management strategy that was enacted as the Biosecurity (National American Foulbrood Pest Management Strategy) Order 1998. An associated order in council called the Biosecurity (American Foulbrood — Apiary and Beekeeper Levy) Order 2003 gave the NBA powers to collect fees, by way of an apiary levy on all beekeepers, to fund the strategy. The NBA raises funds on a voluntary basis for its other activities.

An irrevocable change

In April 2000 the exotic varroa mite was confirmed in Auckland, and beekeeping in New Zealand changed forever. Varroa spread throughout the North Island, and despite attempts to prevent it crossing Cook Strait, the mite was found in Nelson in June 2006. In time varroa will spread throughout the South Island.

Colonies of honey bees with varroa do not survive without beekeeper intervention such as the use of varroa-tolerant bee stock and/or miticides. The time and material costs of managing varroa are significant, and the ratio of hives per commercial labour unit has reduced to an average of around 400. Some beekeepers can manage more hives depending on how many services, such as honey extraction, they contract out.

A brighter future?

Contrary to predictions made when varroa first entered the country, the mite has not led to a significant long-term reduction in beehive numbers. Over 20,000 hives were reported as being killed by the mite in the early years, but as beekeepers learned to manage their hives better and as mite-control chemicals were registered, hive numbers climbed again. There are now more beehives in New Zealand than ever before.

CURRENT STATUS
Beekeepers, apiaries and hives

Table 21.1   New Zealand beekeeping industry (as at June 2010)

Location	Beekeepers	Apiaries	Hives
Northland/Auckland/ Hauraki Plains	618	3331	53,643
Waikato/King Country/ Taupo	196	2316	46,457
Coromandel/ Bay of Plenty/Poverty Bay	296	3493	71,383
Manawatu/Taranaki/ Hawke's Bay/Wairarapa	642	4182	70,143
Marlborough/Nelson/ West Coast	261	1910	28,576
Canterbury	549	3842	54,966
Otago/Southland	395	3366	51,604
New Zealand	2957	22,440	376,772

Source: AsureQuality Limited

Hobby beekeepers with 10 or fewer hives make up 76 per cent of the beekeepers in New Zealand, but manage only 1.6 per cent of the hives. The free-ranging nature of diseases such as varroa and American foulbrood means that hobbyists and inexperienced beekeepers can have an economic impact on other beekeepers. Though hobbyists keep only a tiny fraction of the hives in New Zealand, this sector is an important part of the industry and the source of many new commercial beekeepers.

The great majority of hives are owned by approximately 220 full-time commercial registered beekeeping businesses, with the largest operating over 16,000 hives. The number of full-time and part-time staff employed by commercial beekeepers is probably well in excess of 1000 if office and processing staff are included.

By 2009, apiary and hive numbers recovered after the decline following the varroa incursion in 2000. Most of the increase was driven by new entrants to the industry and commercial beekeepers looking to capitalise on high manuka honey prices. About 200–300 new beekeepers register each year, knowing they have to live with varroa. This bodes well for the future of beekeeping in New Zealand.

Production and value

Table 21.2   New Zealand beekeeping products and services, annual figures

Honey marketed	10,600 tonnes (6-year average; range 8888–12,565 tonnes/year)
Honey exported	4000–6000 tonnes
Beeswax exported	100–150 tonnes
Hives used for commercial pollination	*c*. 200,000
Queen bees exported	3000–10,000
Bulk bees exported	20,000–30,000 x 1 kg packages
Direct sales of hive products and services	$133 million to $150 million

Source: AsureQuality; see also horticulture monitoring: apiculture at www.maf.govt.nz.

INDUSTRY ORGANISATIONS

National Beekeepers' Association (NBA)

The NBA is a voluntary organisation, with approximately 700 members ranging from hobbyist to commercial beekeepers. The executive council is made up of representatives from eight wards (five in the North Island and three in the South). There are 11 branches throughout New Zealand, and members meet at regular intervals to discuss and resolve industry affairs, as well as hold education activities such as field days, workshops and an annual conference.

The NBA is the management agency for the national American foulbrood pest management strategy, and collects a mandatory levy to fund the strategy. The NBA represents its members on all aspects of beekeeping, which may include government policy, research, disease control, hive inspections and education. It publishes a magazine called *The New Zealand Beekeeper*, and maintains a website at www.nba.org.nz.

Fig. 21.4 Sharing experiences at a National Beekeepers' Association field day.

Beekeeping Industry Group (BIG)

The BIG was set up in 2000 as an industry group affiliated with Federated Farmers of New Zealand. It carries out similar functions to the NBA, but most of its members are based in the South Island, particularly in Canterbury. Further information can be accessed through the Federated Farmers' website: www.fedfarm.org.nz

GOVERNMENT AGENCIES AND STATE-OWNED ENTERPRISES

AsureQuality Limited

In 1998 the government reorganised the Ministry of Agriculture and Forestry (MAF). Two state-owned enterprises (SOEs) were formed from part of it to carry out operational activities, while MAF retained core activities such as policy, biosecurity and quarantine. In 2006 the government joined the two SOEs, Asure New Zealand Limited and AgriQuality Limited, back together again under the name AsureQuality Limited. This company provides a range of apiculture services on a contractual basis from offices in Hamilton and Christchurch.

The apiculture group of AsureQuality co-ordinates the American foulbrood strategy for the NBA, and parts of MAF's surveillance and response programme for exotic bee pests and diseases. It provides a verification service to operators who process, handle or store bee products under a risk management programme. It also verifies bee products and live bees for export, audits beehives to pollination standards, participates in a bee product residue monitoring programme, and assists overseas countries to develop or protect their beekeeping industries (www.asurequality.com).

MAF Biosecurity New Zealand

This section of MAF is responsible for market access negotiations and export certification of bees and bee germplasm. It also surveys for, and responds to, the presence of exotic bee pests and diseases (www.biosecurity.govt.nz).

New Zealand Food Safety Authority (NZFSA)

The New Zealand Food Safety Authority is responsible for ensuring the wholesomeness and safety of food offered for sale within New Zealand and also overseas.

If an exporter of a bee product requests an official assurance (export certificate) from NZFSA, either because the importing country requires it or because the certificate may facilitate market access, then that product must have been always processed, handled or stored in premises operating under a risk management programme (RMP). Such programmes are based on the international hazard analysis critical control point (HACCP) system, and the programmes have to be registered with NZFSA and verified at least annually by NZFSA or AsureQuality Limited (www.foodsafety.govt.nz).

On 1 July 2010 NZFSA was merged with the Ministry of Agriculture and Forestry.

Environmental Risk Management Agency (ERMA New Zealand)

ERMA New Zealand administers the Hazardous Substances and New Organisms Act 1996 (HSNO). This Act is designed to protect the environment and the health and safety of communities by preventing or managing the adverse effects of new organisms and hazardous substances such as pesticides. Pesticides were formerly controlled under the Pesticide Regulations 1983 (www.ermanz.govt.nz).

RESEARCH

The New Zealand Institute for Plant and Food Research Limited

This Crown research institute was formed in 2008 from the merger of HortResearch and Crop and Food. It conducts apiculture research, and in doing so follows a long tradition of research by MAF and the former Department of Scientific and Industrial Research. Scientists from Plant and Food have provided valuable information on subjects such as pollination, bee pests and diseases, pesticides, developing bee repellents for the poison baits used in possum control, and a varroa-tolerant bee stock. The main research centre is the Ruakura campus in Hamilton (www.plantandfood.co.nz).

Other Crown research institutes such as Landcare Research and AgResearch also conduct research related to bees, as personnel and funds are available.

Universities

Students and staff at several universities carry out research related to apiculture. Best known perhaps is the research at Waikato University in Hamilton on the antibacterial and anti-inflammatory properties of honey, and new methods of identifying the floral sources of honey (www.waikato.ac.nz and especially bio.waikato.ac.nz/honey).

Otago University also studies honey bees and helped unlock the bee's genetic code. Researchers there are continuing to work on the mode of action of pheromones, and learning and memory in honey bees (www.otago.ac.nz).

Further sources of information

Beekeepers are generally very keen to learn more about their hobby or business, often becoming avid readers or course participants. This chapter will help you to start the search for more information about the honey bee and beekeeping, but there is no need to restrict yourself to the sources listed here. The sky is the limit, thanks especially to the internet.

BEEKEEPING ORGANISATIONS

Join a hobbyist beekeeping club if there is one near you. These clubs can provide a valuable opportunity to learn from others, and you can often share equipment such as honey extractors. Contact details for hobbyist clubs are on the National Beekeepers' Association website and in its magazine.

National Beekeepers' Association (NBA) branches normally provide an environment more suited to commercial or semi-commercial beekeepers, though hobbyists will still find value in their activities, such as field days. Hobbyists can also benefit from the NBA's monthly magazine the *New Zealand Beekeeper*, and the opportunity to borrow books and other information from the NBA's library. For information on all the association's services see www.nba.org.nz

The Beekeeping Industry Group (BIG) provides similar services, but is composed of a smaller number of mainly South Island beekeepers. The BIG is affiliated with Federated Farmers, and has information and a newsletter at www. fedfarm.org.nz/industry/bees.

COURSES

Some training centres hold courses in introductory or advanced beekeeping, which can be a useful way of gaining knowledge and skills. The New Zealand Qualifications Authority (NZQA) offers a National Certificate in Apiculture (Level 2) with 58 credits as an entry-level qualification.

This is for those people working, or considering working, in the industry as assistant beekeepers. The National Certificate in Apiculture (Level 3) is for those working in the industry in a self-directed capacity with levels of unsupervised responsibility (www.nzqa. govt.nz).

Telford Rural Polytechnic (www. telford.ac.nz) provides a 37-week full-time apiculture training programme for those wishing to pursue a career in the apiculture industry or further develop their interest in beekeeping. Students completing this programme can achieve the National Certificate in Apiculture (Level 2) and credits toward higher-level qualifications. Telford also provides study options for a certificate in queen bee rearing, other short courses and correspondence courses.

Agribusiness (www.agribusiness.ac.nz) offers beekeeping courses for beginners at different rural centres, mostly in the South Island.

INTERNET RESOURCES

As it does for any other pursuit, the internet offers a wealth of information on beekeeping. Sites vary in accuracy and how current they are, and as with beekeeping books you need to beware of relying on material written for other countries. But for the discerning user there is a lot of valuable information available. Don't overlook Wikipedia as a useful information source and link to other resources.

Careful use of a search engine will provide you with plenty of material. Here are just a few sites we have found useful. Others are referenced throughout the book.

Extension and research material

Mid-Atlantic Apiculture Research and Extension Consortium (MAAREC)
- maarec.psu.edu/index.html
Useful material on bees and colony management, with a strong focus on pest and disease management

Canadian Association of Professional Apiculturists
- www.capabees.com
Many free extension and research publications

Honey bee research and extension laboratory
- entnemdept.ufl.edu/honeybee/extension/index.shtml
Extension material and self-learning course modules

UK National Bee Unit
- www.fera.defra.gov.uk and follow the links to the National Bee Unit or go to https://secure.csl.gov.uk
Extension material including pdf copies of pamphlets

Beekeeping directories and links

Beehoo
- www.beehoo.com

Beesource
- www.beesource.com

Apiservices
- www.beekeeping.com

Bad beekeeping links
- www.badbeekeeping.com/weblinks.htm

Apitherapy news
- apitherapy.blogspot.com/
Information and updates about apitherapy and other uses of bees and bee products

The site www.extension.org has a beekeeping section, including an interactive 'ask the expert' facility.

MAGAZINES

The *New Zealand Beekeeper* is the monthly magazine of the National Beekeepers' Association. It covers a broad range of topics, with some pages dedicated to the hobby beekeeper. Beekeepers can receive the magazine either by joining the association or subscribing to the journal. Currently the April and October issues of the magazine are sent by the NBA to all registered beekeepers, whether subscribers or not, as there is education or regulatory content in those issues that all beekeepers need to know.

The two major US beekeeping magazines have a good range of articles for the hobbyist and commercial beekeeper — *Bee Culture* (which also publishes a lot of content at www.beeculture.com) and the *American Bee Journal* (www.dadant.com/journal).

The UK *Beekeepers' Quarterly* now incorporates a former magazine called *BeeBiz*. Subscription details and some downloadable issues can be accessed (www.beedata.com). There are also links to a magazine called *Apis-UK* with many past issues available on-line. The *Australasian Beekeeper* has many on-line articles from past issues of this monthly magazine (www.theabk.com.au).

Other magazines and journals are published by the information services International Bee Research Association and Bees for Development.

INFORMATION SERVICES

The International Bee Research Association or IBRA (www.ibra.org.uk) is a non-profit organisation devoted to advancing apicultural science and education worldwide. It provides a comprehensive information network on bees and beekeeping, and promotes beekeeping as a practical form of sustainable development.

IBRA publishes the *Journal of Apicultural Research* (original research papers on a wide range of subjects), now incorporating *Bee World* (topical articles and reviews for beekeepers and scientists). It also publishes the newsletter *Buzz Extra*.

This organisation also has a specialist mail-order publications service, and a unique beekeeping library that is available for members to use by post or in person.

Bees for Development (www.beesfordevelopment.org) provides an information service focused on apiculture in developing countries. It publishes the quarterly *Bees for Development Journal*.

BOOKS

There is a wide range of books on bees and beekeeping, covering hobbyist, commercial and scientific approaches. This list will help you to start a basic beekeeping library, which can be expanded as your interests develop.

Beekeeping books are available from beekeeping equipment stockists or ordered from general online bookstores, and some can even be read or downloaded on-line. Some titles listed might be out of print, but second-hand copies can usually be tracked down at on-line trading sites.

Specialist bee book suppliers include:

- International Bee Research Association: www.ibra.org.uk
- Bee Culture: www.beeculture.com
- Northern Bee Books (UK): www. groovycart.co.uk/cart.php?c=533
- Bees for Development: www.beesfordevelopment.org

Members of the National Beekeepers' Association can also borrow books from its library, and some hobby clubs have their own collection of books.

General beekeeping

Many general beekeeping books are available. Those written in other countries can be useful as a guide, but take care in adapting them to New Zealand conditions. Most, though, are good sources of information about bee biology and basic colony management.

Honey bees and beekeeping: a year in the life of an apiary by Keith Delaplane (University of Georgia, Cooperative Extension Service, Athens, Georgia, 3rd edition, 2006). A 108-page book and video set that is a good source of sound, practical information. www. georgiacenter.uga.edu/tv/videocatalog/ bees.html

The backyard beekeeper: an absolute beginner's guide to keeping bees in your yard and garden by Kim Flottum (Rockport Publishers, 2005). A US guide to beekeeping and to honey and its uses.

The beekeeper's handbook by Diana Sammataro and Alphonse Avitabile (Cornell University Press, Ithaca, New York, 3rd edition, 1998). A clearly written handbook with good diagrammatic explanations.

Insect bites and stings — a guide to prevention and treatment by Harry Riches (International Bee Research Association, Cardiff, 2003). Not specifically about beekeeping, but a useful guide for new or experienced beekeepers, written by a medical doctor who is an experienced beekeeper.

Reference

The first two books are both classic texts descended from first editions produced more than a century ago, and are published by the two major US beekeeping supply firms. They have been comprehensively revised and rewritten.

- *The hive and the honey bee* edited by Joe Graham (Dadant & Sons, Hamilton, Illinois, 1992).

- *The ABC & XYZ of bee culture* edited by H. Shimanuki, Kim Flottum and Ann Harman (A I Root, Medina, Ohio, 41st edition, 2007).

And the ultimate bee book: *Bees and beekeeping: science, practice and world resources* by Eva Crane (Heinemann Newnes, Oxford, 1990). This is an amazingly comprehensive reference book written by a world authority.

Bee biology

The biology of the honey bee by Mark Winston (Harvard University Press, Cambridge, Massachusetts, 1987). A clearly-written and comprehensive introduction to honey bee biology. Despite its age, this is still invaluable for learning about the biological basis for management decisions. Well referenced.

Honey bee biology and beekeeping by Dewey Caron (Wicwas Press, Cheshire, Connecticut, 2000). A good book about honey bee biology and practical beekeeping, for the beginner or intermediate level.

The wisdom of the hive: the social physiology of honey bee colonies by Thomas Seeley (Harvard University Press, Cambridge, Massachusetts, 1995). The result of many years of research, this is a clear account of how this complex animal society works.

The buzz about bees: biology of a superorganism by Jürgen Tautz (Springer, 2008). A comprehensive and beautifully illustrated book covering different themes related to honey bee biology.

Form and function in the honey bee by Lesley Goodman (International Bee Research Association, Cardiff, 2003). A lavishly illustrated guide to the biology of the honey bee, suitable for students, beekeepers and others.

Anatomy and dissection of the honeybee by H. A. Dade (International Bee Research Association, Cardiff, 2009). A reprint of a classic book, illustrated with practical fold-out diagrams.

Volume 57 of the *Fauna of New Zealand* series, by Barry Donovan (Landcare Research, Lincoln, 2007), covers all the native and introduced bees in the country. Available from: www.landcareresearch. co.nz/research/biosystematics/ invertebrates/faunaofnz/extracts/FNZ57/ FNZ57ind.asp

Nectar and pollen sources

Nectar and pollen sources of New Zealand by R. S. Walsh (National Beekeepers' Association, Wellington, revised edition, 1978). The only book on the subject written for New Zealand conditions, detailing over 200 species of native and introduced plants. This is out of print, but is available on the NBA website.

A colour guide to pollen loads of the honey bee by William Kirk (International Bee Research Association, Cardiff, 2nd edition, 2006). Uses new colour recording and reproduction methods to accurately record and help identify pollen loads, with over 500 colours and records of 268 species. Though a UK book, many of the species included are grown in New Zealand.

Pollen identification for beekeepers by Rex Sawyer (Northern Bee Books, Hebden Bridge, 2006). Reprint of a practical guide to identifying pollen types. Describes the characteristics of the main types of pollen found in the UK (but includes many plants also common in New Zealand). The set of punched cards included with the original 1981 edition has been replaced by a CD, sold separately.

Queen rearing

Queen bee: biology, rearing and breeding by David Woodward (Northern Bee Books, Hebden Bridge (UK), 2nd edition, 2009). Written by the former tutor for Telford Polytechnic's queen bee certificate and correspondence courses. Available from Telford.

Queen rearing simplified by Vince Cook (Northern Bee Books, Hebden Bridge, originally published in 1986, reprinted 2004). An excellent handbook that describes very practical methods developed by the author in New Zealand.

Contemporary queen rearing by H. A. Laidlaw (Dadant & Sons, Hamilton, Illinois, revised edition 1979). Though the book is no longer very up to date, it remains a useful and comprehensive guide.

Successful queen rearing manual by Marla Spivak and Gary Reuter (University of Minnesota, Twin Cities, Minnesota, revised edition 2006). Provides practical information on the biology of queens and drones, stock selection and maintenance, and raising queens by the Doolittle method. Companion DVD also available.

Queen rearing and bee breeding by Harry Laidlaw and Robert Page (Wicwas Press, Cheshire, Connecticut, 1997). Aimed at the professional beekeeper.

Pests and diseases

Elimination of American foulbrood disease without the use of drugs – a practical manual for beekeepers by Mark Goodwin (National Beekeepers' Association, Wellington, revised edition 2007). Available from the National Beekeepers' Association. This is the 'bible' on the subject for all New Zealand beekeepers.

Control of varroa — a guide for New Zealand beekeepers by Mark Goodwin and Michelle Taylor (Ministry of Agriculture and Forestry and HortResearch, Wellington, 2008). Available from the National Beekeepers' Association, this revised edition has what you need to know to recognise and control varroa. The book covers experiences of dealing with varroa in New Zealand since 2000, plus the results of research work carried out in this country and overseas.

Diagnosis of common honey bee brood diseases and parasitic mite syndrome by Mark Goodwin and Michelle Taylor (HortResearch, Hamilton, 2003). A useful pamphlet with colour illustrations, which can be downloaded free from: www.hortresearch.co.nz/files/science/ biosecurity/227525-Bee-Pamphletpths-small.pdf

Honey bee pests, predators and diseases by Roger Morse and Kim Flottum (A I Root, Medina, Ohio, revised edition, 1997). Comprehensive coverage of all enemies of beekeeping. North American in perspective but nevertheless has worldwide application.

Honey bee diseases and pests edited by Cynthia Scott-Dupree (Canadian Association of Professional Apiculturists, Guelph, Ontario, 2nd edition, 2000). A practical illustrated guide produced by the Canadian Association of Professional Apiculturists.

Diagnosis of honey bee diseases by Hachiro Shimanuki and David Knox (US Department of Agriculture, Beltsville, Maryland, revised edition, 2000). A US Department of Agriculture handbook on laboratory and field diagnosis, available on-line at: www. ars.usda.gov/is/np/honeybeediseases/ honeybeediseases.pdf

Hive products
Value added products from beekeeping by Rainer Krell (Food and Agriculture Organisation, Rome, 1996). An excellent book with instructions on harvesting and processing bee products, and many recipes. It is available at: www.fao.org/ docrep/w0076E/w0076E00.htm

A book of honey by Eva Crane (Oxford University Press, Oxford, 1980). Covers the production, properties and uses of honey in a very readable way, and despite its age is full of interesting facts and information still relevant today.

Honey in the comb by Eugene Killion (Dadant & Sons, Hamilton, Illinois, 1989 reprint). Originally published in 1981 so

also not a new book, but still the best guide to producing and marketing comb honey. Now sold with an accompanying DVD.

Honey wines and beers by Clara Furness (Northern Bee Books, Hebden Bridge, 2008). This reprint of a classic contains practical recipes and guidance for home winemakers and brewers.

Mead: making, exhibiting and judging by Harry Riches (Northern Bee Books, Hebden Bridge, 1997). Practical recipes and advice from one of the UK's experts in mead-making and judging.

The compleat meadmaker by Ken Schramm (Brewers Publications, Boulder, Colorado, 2003). An authoritative US text.

Beeswax: production, harvesting, processing and products by William Coggshall and Roger Morse (Wicwas Press, Cheshire, Connecticut, 1984). A comprehensive book covering all aspects of this hive product.

How to make beeswax candles by Clara Furness (Northern Bee Books, Mytholmroyd, 2010). A classic guide to candle-making recently reprinted.

Beginner's guide to candle-making by David Constable (Search Press, Tunbridge Wells, 1997). Practical advice and instructions from an experienced candlemaker.

Pollination
Crop pollination by bees by Keith Delaplane and Daniel Mayer (CAB

International, Wallingford, Oxford, 2000). An authoritative but practical research-based guide to using bees for crop pollination.

Insect pollination of cultivated crop plants by Stuart McGregor (US Department of Agriculture, Washington DC, 1976). An older but very comprehensive and relevant guide to crop pollination, now available on-line from *Bee Culture* magazine at www.beeculture.com

A guide to managing bees for crop pollination (1999). A small but useful book from the Canadian Association of Professional Apiculturists, which covers the basics of pollination and how to manage honey bees, alfalfa bees, bumble bees and other bees for successful crop pollination. It is available online from www.capabees.com

New Zealand beekeeping

Bibliography of New Zealand apiculture (1842–1986) by Murray Reid, Andrew Matheson and Grahame Walton (Ministry of Agriculture and Fisheries, Tauranga, 1988). Lists over 1350 references on beekeeping and bees, in subject groupings and with author index. Still available from the International Bee Research Association.

Glossary

Part of the learning curve associated with beekeeping is the specialised terminology that beekeepers use. This can be somewhat intimidating at first, so here we list many of the words and phrases peculiar to the world of beekeeping. You will also find that many terms are defined in the text, and you can locate some of these by referring to the index.

A

absconding swarm
Part or all of the colony that deserts the hive because of unfavourable conditions.

afterswarm
A swarm with a virgin queen that leaves a colony which has already swarmed.

American foulbrood (AFB)
A serious disease of developing honey bees, caused by the bacterium *Paenibacillus larvae larvae. See also Bacillus larvae* (BL).

annual disease return (ADR)
Statutory annual declaration to the AFB pest management agency.

apiarist
Beekeeper or bee farmer.

apiary
A collection of beehives, also the place where bees are kept. Known also as an apiary site (or simply a site) or a bee yard.

apiculture
The keeping of bees, usually for economic benefit.

Apis
The genus to which honey bees belong. *Apis mellifera* is the scientific name for the western honey bee, which is found in New Zealand and is the basis for commercial beekeeping in many countries.

apitherapy
Use of bee products for therapeutic or medical purposes.

B

Bacillus larvae (BL)
The former name for the bacterium that causes American foulbrood, now renamed

Paenibacillus larvae larvae. See also American foulbrood.

bee blower
Electric or petrol-powered fan producing large volumes of air to remove bees from honey supers.

bee bread
Pollen mixed with honey and stored in the comb by honey bees. *See also* pollen.

bee brush
A device for brushing bees from combs; usually honey combs.

bee escape
A device that limits bee movement to one direction and is used to remove bees from honey supers on a hive, or sometimes from cavities such as in a building. Mounted in a bee escape board when used on a hive.

bee product
Honey, honeydew honey, beeswax, venom, propolis, pollen, royal jelly or any other product derived from honey bees.

bee space
A space that honey bees keep open and use as a passage, 6 to 9 mm wide. The discovery of the significance of the bee space led to the development of the movable-frame hive, which laid the foundation for modern beekeeping.

beeswax
A waxy glandular secretion of worker honey bees, used by them to make combs for storing food and rearing brood.

bottom board
The floor or lowest portion of a hive, which usually stands on runners to keep it off the ground.

box
A unit of a beehive that contains the brood (brood box) or the surplus honey (honey box or super).

brace comb
Extra comb built between adjacent parts in a hive and connecting them together. *See also* burr comb.

brood
All stages of developing bees — eggs, larvae and pupae — not yet emerged from their cells.

brood cells or brood comb
Wax cells in which brood is raised.

brood chamber or brood nest
The portion of the hive set aside or used for rearing brood.

brood food
The diet of worker and drone honey bee larvae, consisting of the secretions from hypopharyngeal and mandibular glands of nurse bees. May contain some added honey and pollen. *See also* royal jelly.

burr comb
Surplus wax deposited on a hive part but not connecting it to another. Often 'brace comb' and 'burr comb' are used interchangeably, though their meanings are distinct.

C

capping
A thin sheet of beeswax used by bees to seal a cell containing honey or brood.

caste
A particular form of the honey bee. There are three castes: queen, worker and drone.

cell
A hexagonal (six-sided) compartment of a beeswax comb, used to store food or rear brood.

cell cup
See queen cell cup.

chilled brood
Brood that has died because it has not been kept warm by adult bees in the hive.

chalkbrood
A disease of honey bee brood caused by the fungus *Ascosphaera apis*.

chunk honey
A honey pack containing a piece or pieces of comb honey surrounded by clear liquid honey. *See also* comb honey and cut comb honey.

cluster
The compact group into which bees form themselves inside the hive in cold conditions. Also used to mean a number of bees grouped together, e.g. to build wax combs or when swarming.

cold starvation
Starvation of a colony that still has ample food stores in the hive, but is unable to reach them because extreme cold causes the bees to cluster tightly.

colony
A group of interdependent bees living together.

colony collapse disorder (CCD)
A condition where all, or almost all, the bees suddenly abandon a hive, and the brood and the food stores in the hive.

comb
Pieces of completed wax cells, used for brood-rearing or food storage.

comb foundation
See foundation.

comb honey
Honey complete with the wax comb in which it was stored by the bees. *See also* chunk honey and cut comb honey.

commercial beekeeper
A person whose business is beekeeping, and who derives a significant portion of their income from beekeeping.

creamed honey
Honey that has undergone controlled granulation to produce a firm and finely crystallised honey. *See also* granulated honey, raw honey and starter honey.

crop
See honey sac or honey stomach.

cut comb honey
A type of comb honey in which the comb is cut into portions and placed inside a plastic box. *See also* chunk honey and comb honey.

D

dearth
A period when bees are not able to collect nectar or pollen, due to time of the year, location or weather conditions.

DECA
Disease elimination conformity agreement, a formal agreement between a beekeeper and the AFB pest management agency.

dequeen
To remove the queen from a colony.

division board
A piece of equipment used to divide a hive into two or more parts.

division board feeder
See frame feeder.

drawn comb
A comb made by bees building up cells on the foundation.

drifting
Bees returning to the wrong hive — 'drifting' between hives.

drone
A male honey bee.

drone brood
Developing drones (eggs, larvae or pupae).

drone comb
Comb of cells about 6 mm across, in which the queen bee lays drone eggs.

drone layer
A queen bee that can produce only drone eggs, either because she has never been fertilised or because her sperm supply is exhausted.

E

egg
The first stage of developing brood, from which larvae hatch.

emerging brood
Young adult bees chewing their way out of their brood cells. Sometimes called 'hatching brood', but this more correctly refers to larvae emerging from eggs.

entrance reducer
Piece of equipment used to restrict the size of a hive entrance. Also called a mouse guard.

escape
See bee escape.

European foulbrood (EFB)
A disease of honey bee brood caused by the bacterium *Melissococcus plutonius*.

excluder
See queen excluder.

extracted honey
Honey removed from the comb, usually by spinning in a centrifugal honey extractor.

extractor
A machine for removing honey from the comb.

F

feral colony
A colony of bees living in the wild, not in a beehive. Also called 'wild bees' or 'bush bees'.

field bee
See forager.

fixed-comb hive
A hive without movable frames, in which the combs are stuck to the hive.

flight path or flight lines
Paths used habitually by foraging bees in a colony.

flighty
Bees that are prone to running on the combs and becoming agitated when disturbed.

forager
A worker bee, usually older than a hive bee, which flies from the hive and gathers nectar, pollen, propolis or water.

foulbrood
See American foulbrood and European foulbrood.

foundation
A thin sheet of beeswax imprinted with the hexagonal shape of worker (or in some cases drone) cells. Can also be made from plastic and may or may not be coated with beeswax.

frame
Component of the hive in which foundation is placed for the bees to build combs on. Up to 10 of these hang vertically inside each hive body or box.

frame feeder
A container that hangs in the hive in place of one or more frames, used to feed dry or syrup sugar. Sometimes called a division board feeder.

frass
Excrement and other refuse left by insect larvae, usually wax moths.

fructose
A simple sugar (monosaccharide) found in fruit and honey.

fume board
A top cover with an absorbent pad underneath, which is used with liquid repellents to remove bees from honey supers.

G

glucose
A simple sugar (monosaccharide), the major carbohydrate fuel used by the human body.

grafting
Transferring very young worker larvae from brood cells into artificial queen cells for queen rearing.

grafting tool
A fine brush, hook or scoop used to transfer larvae from worker cells into artificial queen cells.

granulated honey
Honey that has crystallised and become solid. Natural granulation will produce coarse crystals, whereas controlled granulation will produce crystals too fine

to be detected on the tongue. *See also* creamed honey and raw honey.

H

hive
A manufactured container used to house a colony of bees.

hive bee
Young worker bee that performs activities inside the hive and does not forage for nectar or pollen.

hive tool
A metal tool with a flat end, used to prise apart pieces of the hive.

Hoffman frames
A type of self-spacing frame wide enough to provide the correct bee space between combs when the frames are pushed together.

honey
A fluid, viscous or crystallised substance produced by honey bees from the collection and modification of the nectar of flowers or other sources of sugar.

honey flow
A period when nectar is abundant and bees store surplus honey. In reality this is a nectar flow.

honey house
Building used for extracting and processing honey.

honey sac or honey stomach
A compartment in the worker bee's

intestine used for the temporary storage of nectar. Also called the crop.

honeydew
Sugary liquid produced by plant-sucking insects such as aphids and scale insects. May be used by bees as a food source and also transformed into honeydew honey.

hypopharyngeal gland
Gland in the head of worker bees that produces protein-rich components of royal jelly in young workers, and also the enzyme invertase in older, foraging bees.

I

inner cover
Piece of equipment placed under the hive lid to prevent it being stuck to the tops of the frames. Also called hive mat, division board, crown board, top board.

invertase
Enzyme produced in worker bees' hypopharyngeal glands and used to help break down the sucrose in nectar into glucose and fructose.

L

landing board
Short platform in front of the hive entrance for returning bees to land on before entering the hive.

Langstroth hive
A very common type of movable-frame hive named after the inventor of the modern hive (1851). This is the hive type used in New Zealand.

larva (plural larvae)
The second stage of developing bees or brood. Corresponds to the caterpillar or grub stage of other insects.

laying worker
Worker bee with developed ovaries, which is thus capable of laying eggs. These bees never mate and can lay only drone eggs. Found occasionally in queenless hives.

M

management agency
Organisation that manages the American foulbrood pest management strategy.

mandibular gland
Gland in the head of the honey bee. In the queen this gland secretes pheromones responsible for maintaining the social organisation of the colony (queen substance). In young workers secretions from this gland and the hypopharyngeal glands make royal jelly (which is fed to the queen) or worker jelly (which is feed to workers). In old workers of foraging age, the gland also produces heptanone, a component of the alarm pheromone.

Manley frames
Self-spacing frames for honey supers with wide, straight-sided end bars that space the frames at eight per box.

mating flights
Flights taken by a virgin queen soon after she emerges from the cell, during which she mates with drones.

Melissococcus plutonius
The bacterium that causes the bee disease European foulbrood or EFB.

movable-frame hive
Hive in which all the combs are built in frames that can be easily removed from the hive.

N

nectar
Sugary liquid produced by the nectaries of flowers, which are usually located in flowers but can also be on other parts of the plant (where they are called extra-floral nectaries).

nectar flow
A period when nectar is abundant and bees store surplus honey produced from the nectar. Also called a honey flow.

nosema
Disease of adult honey bees caused by an infection of the mid-gut by the protozoan *Nosema apis* and *Nosema ceranae*.

nucleus or nucleus colony (plural nuclei)
A small colony that occupies less than a standard hive box. Also known as a 'nuc'.

nucleus hive
A small box used to house a nucleus colony. Also known as a 'nuc box'.

nurse bee
Young worker bee involved with feeding larvae. A type of hive bee.

O

orientation flights
Flights taken by bees when they first leave the hive, so they learn to recognise the hive's location. Also known as play flights.

overwinter
To maintain a bee colony throughout the winter.

P

Paenibacillus larvae larvae
The bacterium that causes the bee disease American foulbrood or AFB.

paralysis
Viral disease of adult bees that may leave them quivering and unable to walk or fly normally.

pheromone
A chemical produced by one individual that affects the behaviour of another individual of the same species. Much of the honey bee's communication is done with pheromones. *See also* queen substance.

piping
Sound made by queen bees, usually heard when they are in cages waiting to be introduced into beehives.

play flights
See orientation flights.

pollen
The male reproductive substance of flowering plants, which is used by bees as food. *See also* bee bread.

pollen basket
A flattened area on part of the hind legs of worker bees used to carry pellets or balls of pollen grains. The anatomical term for this area is the corbicula (plural corbiculae) — the term 'basket' is commonly used but is not a very accurate description.

pollen substitute
A mixture of protein-rich materials and sugar used to feed bees when natural pollen is not available. Does not contain any pollen.

pollen supplement
A pollen substitute containing natural pollen added to improve its attractiveness and nutritional value.

pollen trap
Piece of hive equipment that removes pollen pellets from the pollen baskets of worker bees as they enter the hive.

pollination
Transfer of pollen from male to female parts of a plant.

proboscis
The mouthparts of a bee.

propolis
Resinous substance collected by bees (usually from plants), then modified and used to seal the inside of the hive and strengthen wax combs.

pupa
The third stage of developing bees, corresponding to the chrysalis stage of some other insects.

pupation
The act of changing from a larva into a pupa. The term is also used for the act of going through the whole pupal stage, which in honey bees occurs in a sealed cell.

Q

queen
Female honey bee with developed ovaries, capable of laying fertilised eggs, that produces pheromones with a profound influence on the colony. The presence and condition of the queen is vitally important to the performance and fate of a colony.

queen cage
Small container usually made of plastic that holds a queen bee, several worker bees and a piece of sugar candy. Used to transport queen bees and introduce them into hives.

queen cell
A large, vertical cell in which a queen is reared.

queen cell cup
The beginning of a natural queen cell, shaped rather like an acorn cup, in which the queen lays an egg. Also an artificial cup used to rear queens.

queen excluder
A screen or mesh with openings large enough to let worker bees through, but which keeps out queens (and drones).

queen raising, queen rearing
The practice of using a colony or colonies to produce queen cells for transfer to other colonies, in order to obtain mated queen bees.

queen substance
A mixture of pheromones produced by the queen, which controls many of the colony's activities. *See also* pheromone.

queenless
The state of a colony with no queen. A colony that has been queenless for some time and has lost the chance to raise a replacement queen is said to be 'hopelessly queenless'. Queenless colonies often make a noise known as a 'queenless roar'.

queenright
The state of a colony with a laying queen present.

R

race
A subspecies of the honey bee, also called breed or strain of bee, e.g. Italian or Carniolan.

raw honey
Honey extracted or scraped from the comb and left to self-granulate. Probably not strained, and can contain pieces of beeswax. *See also* creamed honey and granulated honey.

render wax
To melt down combs and other pieces of wax to recover a block of pure beeswax.

requeen
Replace the queen in a colony with another.

ripe honey (capped honey)
Honey that has been fully ripened and generally capped over. The sucrose has almost all been broken down into the two simple sugars glucose and fructose, and the water content reduced to a level where the honey is stable.

ripen
The process of bees turning nectar into honey by breaking down ('inverting') the sucrose in it into glucose and fructose, and removing much of the water.

riser
Strip of wood around three sides of a hive's bottom board, which lifts up the hive and leaves an entrance for bees.

robbing
The taking of honey from one hive by bees of another, or the frenzied gathering of honey or sugar syrup by bees when there is no nectar flow.

ropey
A diagnostic characteristic of American foulbrood disease, when the decayed larval remains can be withdrawn as an elastic rope or thread.

royal jelly
Food given to queen bee larvae, produced in the hypopharyngeal and mandibular glands of nurse bees. *See also* brood food.

S

sacbrood
A viral disease of honey bee larvae.

scout bees
Worker bees searching for nectar, pollen or water, or for a suitable location for a swarm to settle.

sealed brood
Brood that has been capped over by hive bees, i.e. pupae or old larvae.

section frames
Frames in which section honey is produced.

section honey
Comb honey that is produced in wooden or plastic frames by the bees and sold in that form.

skep
An old-fashioned fixed-comb hive, woven of rushes or straw in the traditional beehive shape.

slumgum
Mixture of pupal cocoons, pollen, propolis and small amounts of beeswax left after brood combs have been rendered down.

smoker
A hand-held metal device used for producing smoke to help calm bees.

split
Part of a colony that is removed to make a nucleus or division, to found a new colony.

split board
A piece of equipment used to divide a hive into two or more parts, usually called a division board.

starter honey
Finely crystallised honey added to liquid honey to start controlled granulation. *See also* creamed honey.

stores
Food for a bee colony — honey and pollen.

summer bees
Worker bees, usually reared in spring or summer, which have low body protein levels and a shorter lifespan. *See also* winter bees.

super
A unit of a beehive that contains frames for storing honey in, placed above or 'superimposed' on the brood chamber. Also known as a box or storey.

supering up
Adding supers to a hive.

supersedure
Process by which a colony naturally replaces an established queen with a daughter queen. Both queens may co-exist for a time.

supersedure cells
Queen cells, usually few in number, of even size and found in the middle of the brood combs. *See also* swarm cells.

swarm
A group of workers and drones that leaves a hive with a queen to establish a new colony elsewhere.

swarm cells
A number of queen cells often found on the bottom of frames and of varying sizes. *See also* supersedure cells.

T

thixotropic
A characteristic of some honeys (such as manuka and ling heather) that makes them set like a jelly in the combs. The honey becomes liquid again if it is agitated.

top
A nucleus that is positioned on top of a hive.

U

uncap
To remove the wax cappings of cells, usually honey cells.

uncapping knife
A broad knife used for removing cappings from honey combs before extraction, usually heated by hot water, steam or electricity.

uniting
Joining two colonies into one.

unsealed brood
Brood that has not been capped over, i.e. eggs and larvae.

V

varroa
A parasitic mite of honey bees, known scientifically as *Varroa destructor*.

virgin
A queen bee that has not mated.

W

wax
See beeswax.

wax glands
Four pairs of glands on the stomach or abdomen of the worker bee that secrete beeswax.

wax moths
Species of moth whose larvae damage beeswax combs by tunnelling through them and eating the pollen, wax, honey and debris.

wild bees
See feral colony.

winter bees
Worker bees, usually reared in autumn, with a high body protein level and a long lifespan. *See also* summer bees.

wired frames
Frames with several lengths of wire stretched across, ready for foundation to be fastened in.

worker bees
Female bees with undeveloped ovaries, which perform most of the functions of the colony.

worker comb
Comb containing cells approximately 5 mm across in which the queen normally lays worker eggs. Also used for the storage of honey and pollen.

Handy tips

How much, how many, what, where? There are a lot of practical questions in beekeeping, and here are some of the answers we have gathered during our beekeeping careers from our own experience, and from observations made by other people.

You might have a different view on some of these — sometimes there are no right answers — but we hope that these tips give you at least a starting point.

Bees: number to weight ratio
Newly emerged: 9900/kg.
Foragers: 11,000/kg.
Swarming bees (which have fuller honey stomachs): 7700/kg.

Comb foundation
These are just some of the options — see manufacturers and stockists for a full range.
Note that there are variations between suppliers.

Grade	Size	Sheets /kg
Thin super	Full depth	24–26
Thin super	¾ depth	35–37
Thin super	Half depth	48.5–56
Medium brood	Full depth	17–17.5
Medium brood	¾ depth	21–23
Heavy brood	Full depth	13–15
Drone comb	Full depth	9

Cut comb honey processing

Commercial producers report being able to cut and pack about
100 dozen per day (3.5 people) and up to 250 dozen per day (six people).

Candy for queen cages

Mix 2.3 kg of granulated sugar, one litre of water and a pinch of tartaric
acid (a quarter of a teaspoon). Bring to a boil and simmer very slowly for
30 minutes.

Allow this syrup to cool, but while it is still warm mix it with piping
sugar (one part syrup to five parts piping sugar). Add three or four drops
of glycerine.

Knead with extra piping sugar to achieve a soft consistency.
The final candy should be somewhat softer than the equivalent honey
candy. Store until needed in airtight plastic bags in a refrigerator.

Note that piping sugar is a non-starch icing sugar available from some
cake decorators or bakers. Some beekeepers make their own starch-free
icing sugar by grinding ordinary white table sugar in a kitchen blender
until it is powdered.

Some people say that icing sugar containing starch should never
be used for making candy, but others report using it for years with no
apparent ill-effects on the bees.

Hive equipment: assembling frames

With hammer, nails and glue, and six nails per frame, one person averages
37 frames an hour.

With the use of an assembly box this goes up to 48 per hour.

With an assembly box, glue and staple gun, a person can assemble
72 an hour.

Further mechanisation or leaving out the glue can increase the speed
even more.

Hive equipment: assembly

Combined assembly time: assembling and painting one box, assembling,
wiring and inserting wax foundation into 10 frames, total 55–60 minutes.

Composite time for assembling and painting one hive is about eight
hours. This includes four boxes, lid, floor, excluder, inner cover and
frames (nailed, wired and waxed).

Hive supers: buying timber

A full-depth box uses approximately two linear metres of timber. A cubic metre of 240 mm x 20 mm dressed timber will yield around 160 linear metres and make 80 full-depth boxes. The final number depends a bit on the grade of timber and how many knots are cut to waste. Premium grade is usually knot-free, and commercial or merchant grade will have tight knots. Packing-grade timber will have a quite a lot of knots, some of which may fall out over time as the timber expands and contracts with the weather.

Hive equipment: assembly: nails

Size	Type	Purpose	Approx Number/kg
12 x 1.0 mm	Vinyl or cement/ glue coated	Wiring tacks	12,000
30 x 1.6 mm	Vinyl or cement/ glue coated	Frame assembly	2000
40 x 1.6 mm	Galvanised	Supers, roofs and risers on floors	930
50 x 2.5 mm	Galvanised	Supers, roofs and floors	440
60 x 2.5 mm	Galvanised	Supers, roofs and floors	360

Most commercial beekeepers use staple guns and/or Posidriv® screws to assemble supers.

Hive equipment: preservation: paraffin wax

Use paraffin wax with a melting point of 60–62 °C. This has a density of 870–890 kg/cubic metre, and a specific heat of 0.69.
When dipping new boxes, four boxes will consume 1 kg of wax.
For previously-dipped boxes, plan on around six boxes per kilogram.

Hive equipment: preservation: painting

When painting paraffin-dipped full-depth boxes that have previously been dipped and painted, using two coats of water-based paint applied to hot boxes you can expect one litre of paint to cover around 16 full-depth boxes.

Honey supers: weight

Super	Frames	Full (kg)	Extracted wet (kg)	Extracted dry (kg)	Foundation (kg)
Three-quarter depth	8 Manley	27.5	7.0	6.2	5.9
Full-depth	8 full Hoffman	39.3	9.7	8.8	8.4

Empty pine supers (without frames) weigh 4.6 kg.

Hive equipment: wiring frames

Wiring frames with a wiring board takes about one minute per frame once you are in the swing of it. Expect to do five boxes per hour (10 frames) if the end bars have four holes, and six boxes per hour with three-hole end bars. A 200 g reel of 28-gauge frame wire (0.46 mm in diameter) will wire about 100 full-depth frames with three-hole end bars, so a 2 kg reel will wire about 1000 frames.

Honey processing

Nylon honey strainers (280 microns fine, 425 microns medium to coarse).

Pollen: building traps

The trapping mesh has 4.28 mm square apertures (five mesh to the inch woven wire), with 0.90 mm diameter wire. The mesh over the pollen trays is six-mesh to the inch woven wire (3.53 mm square apertures x 0.71 mm diameter wire). This mesh prevents bees getting into the pollen in the collection tray. Perforated plastic sheets can also be used to strip pollen off the bees' legs. The ideal hole size is 13/64 of an inch — if you still have an old set of drill bits. The metric equivalent is 5 mm, which will work but is a little bit on the small size.

Pollen: drying

Bee-collected pollen can contain over 20 per cent water, and should be dried to between six and eight per cent. It should be dried at 36–38 °C, certainly not more than 45 °C, and out of direct sunlight as this can bleach the colour out of some pollens. It should dry within 24 hours, but this depends on the ambient humidity and the nature of your drying equipment. Dehumidifiers can speed up the drying process.

Royal jelly yields

In one trial the most royal jelly was obtained by harvesting cells at three days after grafting (235 mg/cell, or 4255 cells to obtain 1 kg of royal jelly). You can produce about 500 g of royal jelly per hive over a three-month season, with a labour input of about 0.5 h/hive/day. This will vary in different places and in different seasons. Royal jelly can be stored at 2 °C for a year, or for several years either at –18 °C or by freeze-drying.

Wax processing: cappings spinners

The relative centrifugal force (RCF) can be calculated by:

$$RCF = \frac{r \times N \times N}{90,000}$$

Where   r   =   the radius in mm
(distance from the central axis to the surface where the cappings sit)

N   =   the rotation speed in revolutions per minute

For cappings spinners an RCF of around 20 (i.e. 20 times the force of gravity) is common, as this is regarded as being most efficient at removing honey from wax.

Wax processing: solar melters

Placement of material	Efficiency of wax recovery
Old combs, several layers deep	About 33%
Old combs, one layer deep	50%
New combs	80%

Syrup feeding tables

Table 1: The weight of sugar used to produce a particular volume of syrup of different strengths, and the weight of resulting food (honey) stored by a colony.

Sugar used	Sugar to water ratio (by weight)								
	1:1			1:1.5			2:1		
(kg)	Water litres	Syrup litres	Stores kg	Water litres	Syrup litres	Stores kg	Water litres	Syrup litres	Stores kg
1.0	1.0	1.6	1.0	0.67	1.27	1.12	0.5	1.1	1.2

For example, mixing 1 kg of sugar with one litre (which is 1 kg) of water produces 1.6 litres of syrup, from which a colony will store 1 kg of honey. Mixing 1 kg of sugar with 0.67 litres of water produces 1.27 litres of syrup and 1.12 kg of stored honey, and so on.

Table 2: The volume of syrup resulting from various proportions of sugar and water, and the weight of resulting food (honey) stored by a colony.

Volume of syrup	Sugar to water ratio (by weight)								
	1:1			1:1.5			2:1		
(litre)	Sugar kg	Water litres	Stores kg	Sugar kg	Water litres	Stores kg	Sugar kg	Water litres	Stores kg
1.0	0.62	0.62	0.62	0.79	0.53	0.90	0.91	0.45	1.09

For example, to get one litre of 1:1 syrup mix 0.62 kg of sugar with 0.62 litres (or kg) of water. Coincidentally, this will turn into 0.62 kg of honey stores. At the other end of the table, to get one litre of 2:1 syrup, mix 0.91 kg of sugar with 0.45 litres (kg) of water. This will get a colony to store 1.09 kg of honey, showing that heavy syrup is better suited to increasing a colony's honey stores (and lighter syrup is better at stimulating brood-rearing).

Table 3: The amount of stored food obtained by feeding syrup made from various proportions of sugar and water.

Stored food	Sugar to water ratio (by weight)								
	1:1			1:1.5			2:1		
(kg)	Sugar kg	Water litres	Syrup litres	Sugar kg	Water litres	Syrup litres	Sugar kg	Water litres	Syrup litres
1.0	1.1	1.1	1.6	0.89	0.59	1.13	0.84	0.42	0.92

For example, to obtain 1 kg of additional stores in a hive by using 1:1 syrup, add 1.1 kg of sugar to 1.1 litres of water, which will make 1.6 litres of syrup. At the other end of the table, to get 1 kg of stores from 2:1 syrup you need only 0.84 kg of sugar (mixed with 0.42 litres of water, which will make 0.92 litres of syrup). Again, this shows that heavy syrup is better suited to increasing a colony's honey stores. It also shows that a lot of sugar is needed to feed bees, as a strong colony can consume a frame of stored food (which is more than 1 kg) in a day or so.

Wax processing: steam rendering

Ten combs built on medium brood foundation yield 1.6 to 2.0 kg
(average 1.7 kg) of beeswax when melted out in a steam chest.
Few beekeepers bother with further melting down slumgum, the waste
product from a steam chest. In one trial, 19.1 kg of slumgum yielded
only 1.2 kg of dark wax, or six per cent by weight.
Most beekeepers just spread slumgum on the garden, where it makes
an excellent mulch and fertiliser.

Wax yield: cappings

Each tonne of honey produces 14–18 kg of refined cappings wax,
depending on the methods of uncapping and wax recovery.
When cappings are spun out in a spinner about 1.25 to 2 per cent of
the honey crop is still bound up in the 'dry' cappings.
Modern spin float centrifuges and presses may extract more honey.

Index

Italic numbers refer to illustrations, and where there is more than one reference any principal reference is given in **bold**.

Some topics have not been individually indexed but can be found easily by looking at the relevant chapters: individual plant species referred to as nectar or pollen sources, or crops for pollination; books, magazines and other information sources; entries in the glossary; people listed in the acknowledgements.

absconding swarm *see* swarm
access to apiaries 22–3
Achroia grisella see wax moth
acquiring bees *see* bees, acquiring
adult bee diseases *see* nosema disease, paralysis
afterswarm *see* swarm
AgResearch 252
Agribusiness 254
air freshener, use in colony management 119, 134
alkali bee 63
allergic reaction to stings 8
Alley method of queen rearing 137–8, *137*
American foulbrood **156–66**, 172–3
    beekeeping management to reduce and eliminate **163–5**, 187
    causative agent 156
    checking colonies for 155
    cleaning equipment contaminated with 161–2, *161*
    destroying infected hives 160–1, *161*
    development in a colony 162
    historical impact 244
    pest management strategy 234–6
    regulations concerning 165–6, 234–6
    spread 162–3

sterilising contaminated hive equipment 161–2, *161*
symptoms 157–9, *157, 158, 159*
Amitraz *see* Apivar
Apiaries Act *see* legislation
apiary
    access to 22–3
    arranging hives within 23–4
    registering 24, **233–4**
    selecting the site for 21–3
Apiguard (thymol) 149
Apilife VAR 149, *149*, 151
Apis cerana see Asian honey bee
Apis mellifera see honey bee
Apistan (fluvalinate) 144, 149, 151
    resistance to 144–5
Apivar 149, 151
Ascosphaera apis see chalkbrood
Asian honey bee 45, 145
AsureQuality 251
Australian paper wasp *see* wasp

Bacillus larvae *see* American foulbrood
baffle tank *see* honey tank
Bayvarol (flumethrin) 149, 151
bee blower 191–2, *192*
bee brush 188
    harvesting honey with 188–9, *188, 189*
bee dance 56–7

bee escape and escape boards 189–90, *190*
bee, native 62–3
bee products
    importing 236
    types of *see* beeswax, honey, pollen, propolis, royal jelly
Bee-Quick 191
bee space 27, *27*
bee veil *see* veil
beech (*Nothofagus* species) as honeydew source 194, *194*
beehive *see* hive
beekeeping
    clubs 10, 253
    courses 253
    deciding whether to keep bees 8–10
    starting with bees 10–11, **12–26**, 36
    statistics 249–50
    see also colony management
Beekeeping Industry Group 242, 251
bees, acquiring 14–20
beeswax 217–25, *224*
    cappings 218, *218*, 219–22, 279
    culled combs 218–19, *220*
    hot-water melter 223–4, *223*
    polish 225
    processing **219–24**, 277, 279
    properties 217
    scrapings 218, *219*

selling 224
solar wax melter 221–2, *221*, 223, 227
sources 217–18
uses 224
Beltsville bee diet 106–7
benzaldehyde 191
biology of the honey bee 45–62
Biosecurity Act (1993) *see* legislation
black honey bee *see* European race of honey bee
blower *see* bee blower
Boardman feeder *see* entrance feeder
Bombus species *see* bumble bees
boots 13–14
bottom board *see* floorboard
box hive *see* hive, fixed comb
boxes *see* hive box
brace comb 35
Bray, W B 244
brood *155, 156*
    caring for by workers 52
    development *46, 47*–51 *see also* egg, larva, pupa
    inspection for disease identification 155
brood boxes 29–30
brood cells 48–51
brood chamber 29, 101, 111
brood diseases and disorders *see* diseases and disorders of brood
brood food 50–1
brood nest *see* brood chamber
brush, bee *see* bee brush
bumble bees 64
Bumby, Mary 243
burr comb 218
Butler, Charles 125

camphor 151
candy for queen cages 274
cappings *see* beeswax
carbaryl for wasp control 182, 183
Carniolan race of honey bee 25, 45, 112, 243, 244
castes of honey bee 46
    developmental stages 47
    see also drone, queen and worker honey bees
chalkbrood **167–9**, 172–3
    control 169

prevention 169
spread 168–9
symptoms 168, *168*, 169
chemicals *see* mite-control chemicals
chilling of adults 175
chunk honey *see* comb honey
clothing *see* protective equipment
clustering by a bee colony 95, *95*
cold starvation 95
Colletidae 62–3
colony collapse disorder 176
colony management 83–97
    autumn 92
    before the honey flow 86–7
    dividing colonies 115–18
    honey flow management 87–9
    moving *see* hives, moving
    pollination hive management 91–2
    seasonal management plan 96–7
    spring 85–6
    uniting colonies 118–19, *119*
    winter 95
    wintering down 92–4
    see also feeding bee colonies
comb
    building by worker bees 53–4, *53*
    culling 94, *94*
    natural *53*
comb foundation *see* foundation
comb honey 211–16
    chunk honey 216
    colony management for producing 213–14
    cut comb honey 211–12, 215, *215*
    equipment for producing 211–13
    processing and packaging 214–16, 274
    section honey **212–13**, *212, 213*, 216
communication by bees 56–7
Coriaria see tutu
Corynocarpus laevigatus see karaka nectar
Cotton, William 243
cut comb honey *see* comb honey

diabetes and honey 198
diseases and disorders of brood 155–73
    inspection for 155

see also American foulbrood, chalkbrood, European foulbrood, half-moon syndrome, parasitic mite syndrome, sacbrood
dividing colonies *see* colony management
drifting
    definition 23
    reducing 24, *24*, 117
drone honey bee *46*, **49–50**
drugs for bee disease control 236
dysentery 175

economics of beekeeping 9, 247
egg *46*, 47–51
embedding foundation 39, *39*
emergency queen rearing impulse 48
entrance feeder *see* feeder, entrance
Environmental Risk Management Authority 241, 252
escape board *see* bee escape
eucalyptol 151
European foulbrood **169–71**, 172–3
    symptoms 170–1, *170*
European race of honey bee (black bee) 45, 243
excluder *see* queen excluder
extractor *see* honey extracting
eyesight 9

fanning 55
feeder
    entrance (Boardman) feeder 103–4, *104*
    frame feeder 102, *102*
    inverted container 102–3, *103*
    top (Miller) feeder 104, *104*
feeding bee colonies
    autumn 98–9
    honey 100–1, *100*, 102
    nucleus colonies 15–16
    pollen 106
    pollen substitute 106–7, *107*
    quantities 99–100
    spring 98
    sugar, dry 101, 105, *105*
    sugar syrup 101, **102–5**, *102, 103, 104*, 278

water 25, 106
winter preparation 93
see also feeder
feral colony
transferring to a hive *18, 19*–20
floorboard (bottom board) 34–5, *34*
tunnel entrance *35*
ventilated (mesh) floorboard **34–5**, 123
flumethrin *see* Bayvarol
fluvalinate *see* Apistan
food sources for bees 22
see also feeding bee colonies
foraging behaviour **58–9**, 61–2
formic acid 149, 150, 152
foundation 32–3, *89*, 224
embedding 39, *39*
grades 273
plastic 89, *89*
purchasing 224
storage 224
use in honey supers 89–90
frame
assembling 37, *37*, 274
Hoffman *30*, 31, *31*
Manley *30*, 31–2
plastic 32, 89
rebates *30, 30*
Simplicity 32
wiring **37–8**, 276
wooden 31–2
frame wiring board 37–8, *38*
fumagillin (Fumidil B) 174, 236
fume board 190–1, *190*
fungicides 40–1

Gibb, R 244
gloves 14
using gloves 25
guarding behaviour 54, *54*

half-moon syndrome **171**, 172–3
Halictidae 63
handling bees 65–71
hanging out 88, *88*
history of New Zealand beekeeping 243–8
hive
acquiring 16
bee space 27, *27*
fixed-comb types 27, 244

historical hive types **27–8**, 243–4
inspecting 67–70, *68, 69, 70*
locating 21
movable-frame, first use 244
see also floorboard, foundation, frames, hive box, hive lid, hive mat, queen excluder
hive barrow 123–4, *123*
hive box 29–30
assembling 36–7
timber for 273
weight 274
hive entrance
entrance reducer 94
tunnel entrance *94*
hive equipment 27–44, *29*
assembling **36–40**, 274
dimensions 28
importing 236
nails 275
painting 275
preserving **40–4**, *42*, 275
sources 29
timber types 29
see also floorboard, foundation, frame (wooden and plastic), hive box, hive lid, hive mat, queen excluder
hive lid
assembling 39–40
flush 33–34, *33*
migratory 34, *34*
telescopic 33, *33*
types 33–4
hive loader 124, *124*
hive management *see* colony management
hive mat 35–6
assembling 40
hive tool 14
hives, moving 25, **120–4**, *122, 123, 124*
honey
acidity 197–8
antibacterial properties 199–200
calorific value 198
colour 197
comb *see* comb honey
composition 194–5
creamed *see* honey, granulation and granulating

definition 193
extracting 201–7
granulation and granulating 198, **208–9**
harvesting 186–92
liquid 207
marketing 210, *210*
processing **207–9**, 276
properties 195–200
refractive index 196–7
sources **72–82**, 193–4
tank 206–7, *206*
thermal conductivity 198
thixotropy 196
uncapping 202–4, *203*
viscosity 195–6, **202**
water content 196–7
honey bee 45
Asian honey bee (*Apis cerana*) 45
behaviour 52–62
biology 45–62
imports to New Zealand 45, 236, 243–4
weight 271
honey box (super)
adding to hive **88–9**, 111
bottom supering 89
top supering 88
weight 274
honey extractor 204–5, *205*
Honey Marketing Authority 246–7
honey tank 205–7, *206*
honeydew
beech 193–4, *194*
definition 193
tutu 194,
Hopkins, Isaac 244
Hymenoptera 62

importation of honey bees *see* honey bees, imports to New Zealand
inner cover
assembling 40
insecticide poisoning 175, 229–30, *230*, 241–2
International Bee Research Association 255, 260
Israeli bee paralysis virus 176
Italian race of honey bee 25, 45, 243

karaka nectar 175–6
kowhai nectar 176

lactalbumin 106–7
Landcare Research 252
landing board *see* floorboard
Langstroth hive 27–8
Langstroth, L L 27
larva 46–9, *46*, 51, 52
laying worker *see* worker
honey bee
laws *see* legislation
leafcutter bee *see* lucerne
leafcutter bee
legislation 10
  Agricultural Compounds and
    Veterinary Medicines Act
    (1997) 236
  Animal Products Act (1999)
    232, 237
  Animal Products (Harvest
    Statement and Tutin
    Requirements for Export Bee
    Products) Notice (2009) 232,
    240
  Animal Products (Regulated
    Control Scheme – Verification
    of Contaminants in Bee
    Products for Export) Notice
    (2009) 238
  Animal Products (Specifications
    for Products Intended
    for Human Consumption)
    Amendment Notice (2009)
    238
  Apiaries Act (1907, 1969) 234,
    244, 245, 248
  Australia New Zealand Food
    Standards Code (2002) 238
  Biosecurity Act (1993) 232,
    236, 248
  Biosecurity (American
    Foulbrood — Apiary and
    Beekeeper Levy) Order (2003)
    232, 234, 235–6, 248
  Biosecurity (National American
    Foulbrood Pest Management
    Strategy) Order (1998) 232,
    233–4, 248
  Consumer Guarantees Act (1993)
    238
  Dietary Supplement Regulations
    (1985) 241

drugs for disease control 236
export assurance 237–8
export control, historical 245–7
Fair Trading Act (1986) 239
Food Act (1981) 232, 237, 241
food control programme 237
Food Hygiene Regulations
  (1974) 232, 237
food safety programme 237
Food (Tutin in Honey) Standard
  (2008) 232
Food (Tutin in Honey) Standard
  (2010) 239–40
Hazardous Substances and New
  Organisms Act (1996) 241–2
Health Act (1956) 233
importation of bees, bee products
  and beekeeping equipment
  236
labelling 238
Local Government Act (2002)
  233
medicinal claims 241
Medicines Act (1981) 241
nutraceuticals 241
pesticides 241–2
residue testing 238
risk management programme
  237–8
Weights and Measures Act
  (1987) 239
lucerne leafcutter bee 63–4, *63*

mason bee *see* red clover mason
  bee
medicinal claims for food 241
Megachile rotundata see lucerne
  leafcutter bee
Megachilidae 63–4
Melisscoccus plutonius see
  European foulbrood
menthol 151
mice 185
  damage *185*
Miller, C C 110–11
mineral oil 149, 152
Ministry of Agriculture and
  Forestry
  MAF Biosecurity New Zealand
    236, 251
Ministry of Agriculture and
  Fisheries 247
New Zealand Food Safety

Authority 233, 236, 237, 238,
  239, 241, 251
mite-control chemicals 8–9
  effect on queens 143
  organic chemicals 150
  residues 150
  resistance to 144–5, 151, **152–4**
  synthetic chemicals 151
  see also Apilife VAR, Apistan,
    Apivar, Bayvarol, camphor,
    fluvalinate, formic acid,
    mineral oil, Mitegone, oxalic
    acid, thymol, Thymovar
Mitegone 152
moving hives *see* hives, moving

naphthalene 180
narcotic nectar *see* kowhai nectar
Nasonov gland 55
National Beekeepers' Association
  of New Zealand 145, 234, 242,
    247, 250, 253
native bees 62–3
nectar
  collection by worker bees 59, *59*
  ripening by worker bees 60–1
  sources 72–82
neighbours *see* public relations
neonicotinoids 176
nest cleaning by worker bees 52
New Zealand Food Safety
  Authority *see* Ministry of
  Agriculture and Forestry
New Zealand Institute for Plant
  and Food Research 252
New Zealand Qualifications
  Authority 253
newspaper method of uniting
  colonies 118–19, *119*
Nomia melanderi (alkali bee) 63
Nosema apis see nosema disease
Nosema ceranae 176
  see also nosema disease
nosema disease 174, 236
nucleus colony *116, 131*
  acquiring 14–15
  installing 15–16, *15*
  introducing queen cell to 142–3
  making 115–17
  use for swarm prevention 112
nutraceuticals 241

orientation flights 55–6
Osmia coerulescens see red clover
    mason bee
Otago University 252
overalls 13
oxalic acid 149, 150, 152

package bees
    acquiring 20
    installing 20
*Paenibacillus larvae larvae
    see* American foulbrood
paint and painting of hive
    equipment 41, 43, *43*, 275
paper wasp *see* wasp, paper
paradichlorobenzene 179–80
paraffin dipping of hive equipment
    41–4, *42*, *43*, *275*
paralysis 174–5
parasitic mite syndrome *see*
    varroa
passion vine hopper 194, **239**, *239*
permission of landowner for apiary
    site 21–2
Plant & Food *see* New Zealand
    Institute for Plant and Food
    Research
plywood 44
Polistes humilis and *Polistes
    chinensis see* wasp, paper
pollen
    collection by worker bees 58–9,
        59
    drying 276–7
    sources 72–82
    trapping 276
pollination *91*, **226–31**
    colony management for **91–2**,
        229
    colony stocking rate for 230–1
    honey bees as pollinators 226–7,
        227
    using beehives for 228–9
preservation *see* hive equipment,
    preserving
propolis
    collection by worker bees 61–2,
        62
protective equipment 12–14, *13*
    see also boots, gloves, hive tool,
        overalls, smoker, veil
public relations 24–6
pupa *46*, *47–9*, 51

pyrethroid insecticides for wasp
    control 183

queen cell
    buying 131
    introducing 142–3
    protecting 142–3
    transporting 142
    using 141–3
queen excluder 36, *90*,
    assembling 40
    use 90–1
queen honey bee *46*, **46–9**, *48*,
    125, *125*
    buying 131
    candy for queen cages 272
    caring for by beekeepers
        131–2
    caring for by worker bees
        51, 53
    development 47–8
    egg-laying 49
    finding 125–7, *126*
    introducing 132–4
    marking 127–8
    mating 48
    problems with 128
    transporting 132
    see also queen rearing,
        requeening
queen rearing 134–43
    Alley method 137–8, *137*
    grafting methods 138–41, *139*,
        140
    Miller method 136–7, *137*
    natural queen-rearing impulses
        47–8, 135
    non-grafting methods 136–8
queen substance 49, 132

races of bee *see* Carniolan,
    European, Italian races of
        honey bee
red clover mason bee 64
regulations *see* legislation
requeening
    importance of 111, **128–9**
    timing 130
riser *see* floorboard
robbing by worker bees 60, *60*
    preventing 15, **187–8**
royal jelly, yields 277

Ruakura 245
runner *see* floorboard

sacbrood **166–7**, 172–3
    symptoms 166–7, *166*, *167*
    treatment 167
scenting 54–5, *55*
Scolypopa australis see passion
    vine hopper
scrapings *see* beeswax, scrapings
section honey *see* comb honey
shelter for hives 22
smoker 12
    lighting and using **65–7**, *66*, 126
Sophora microphylla and *Sophora
    tetraptera see* kowhai nectar
Sphecophaga see wasp
starting beekeeping 10–11, **12–26**
stings and stinging 70–1, *70*, *71*
    allergic reaction to 8
summer bees 51
sunlight in apiary site 22
super, supering *see* honey box
supersedure of queens 47
swarm *17*, 110
    absconding 108
    capturing 17–18
    cells 114
    control 112–14
    hiving 18, *18*
    prevention 110–12
    ten-second swarm cell check
        113, *113*, 114
swarming 47–8, **108–10**

Tanalised timber 44, 175
Tank *see* honey tank
Telford Rural Polytechnic 254
thixotropic (honey) 196, 205
thymol 149, 151
Thymovar 149
timber for hive construction 29,
    275
timing, importance of in colony
    management **9**, 65
top board, top mat *see* hive mat
toxic honey 194, **239–40**
toxic nectar *see* karaka nectar
toxic products in timber
    preservation 44, 175
tutu as source of toxic honeydew
    194, 239–40

uncapping knife 202–4, *203, 218*
uniting colonies *see* colony
  management

varroa 144–54, *145, 146, 147*
  chemical treatment for 149–52,
    149, 151
  discovery in New Zealand 144,
    248
  effect on bees and colonies 147
  life cycle 146
  life span 146
  management lessons for 154
  parasitic mite syndrome **147–8**,
    148, 172–3
  regulations 236
  resistance to chemical treatment
    152–4
  sampling hives for 148–9, *149*
  treating hives for 149–52, *149,
    151*

  see also mite control chemicals
Varroa destructor see varroa
Varroa jacobsoni see varroa
veil 12–13
ventilation in hives 94, **111–12**,
    122, *122*
Vespula germanica and *Vespula
  vulgaris see* wasp
Von Frisch, Karl 56, 57

Waerenga queen rearing station
    245
Waikato University 252
Wakefield, Arthur 243
wasp, common and German *181,*
    181–3
  control 182–3
  parasite 181
wasp, paper 184–5
water
  collection by bees 61, *61*

providing for bees 25, 106
wax *see* beeswax
wax moth, greater and lesser *176,*
    177–81, *178*
  control 178–81
  damage 178–81, *177*
  life cycle 177–8
weight of boxes and hives 9, 276
white waxing 87, *87*
wild colony *see* feral colony
winter bees 51
wintering down *see* colony
    management
wiring board for frames, 37–8, *38*
worker honey bee *46,* **51***, 51*
  development of *46,* 47
  laying worker *50*
  life span 51

yeast, brewer's 106–7